AMERICAN CHEMICAL ENTERPRISE

A Perspective on 100 Years of Innovation

To Commemorate the Centennial of the

Society of Chemical Industry (American Section)

BY
MARY ELLEN BOWDEN
AND
JOHN KENLY SMITH

CHEMICAL HERITAGE FOUNDATION
PHILADELPHIA
1994

CHEMICAL
HERITAGE
FOUNDATION

PUBLICATION
NO. 14

ISBN #0-941901-13-0

1994 by the
Chemical Heritage Foundation

Design by Sylvia Barkan
Composition by Pat Wieland
and Ruttle, Shaw & Wetherill
Printing by The Sheridan Press
Production by Frances Kohler

For further information,
including copies of *American Chemical Enterprise*,
please contact the

Chemical Heritage Foundation
3401 Walnut Street, Suite 460B
Philadelphia, PA 19104-6228 USA

Telephone (215) 898-4896
Fax (215) 898-3327

Contents

5   Foreword by Harold A. Sorgenti

7   Introduction
Entrepreneurship and the Promise of America

16   The Spark of Genius
Electrochemicals and Electrical Inventions

23   Earth and Air
Mineral Wealth and Industrial Gases

31   Mauve and More
From Natural Products to Synthetic Organic Chemicals

40   Mortarboard and Lab Coat
Chemical Education and Industrial Research

54   Cracking Nature's Secrets
Petroleum and Petrochemicals

62   A Symphony of Synthetics
Nylon, Synthetic Rubber, and Plastics

69   Miracles from Molecules
Pharmaceuticals

78   In a Global Village
The Environment and the Information Age

87   In Conclusion
From Entrepreneurs to Managers to Statesmen

93   American Section
Chairmen; Perkin Medalists; Chemical Industry Medalists

96   For Further Reading

The banquet held by the American Section of the Society of Chemical Industry, on 6 October 1906 at Delmonico's in New York, to celebrate the fiftieth anniversary of William Henry Perkin's discovery of mauve, or aniline purple, the first coal-tar dye (see pages 13–15). Chemical Heritage Foundation collection.

Foreword

The mind of the nation flung itself into the mighty prospect, dreamed for decades of the comforts that we now take for granted, and positively lusted for the chance to yield to the gratifications of technology.

—Perry Miller, "The Life of the Mind
in an Age of Machines" (1961)

The story of the American Section of the Society of Chemical Industry (SCI) coincides with the growth of the chemical sciences and chemical process industries during the twentieth century. For our centennial, we have chosen to tell that story through the achievements of the holders of the Perkin Medal and the Chemical Industry Medal. If written for SCI's fiftieth anniversary celebration, our tale might have been called a "romance." The word evokes the delight that Americans took in hearing of people who use their creative talents imaginatively to invent, to innovate, and to organize new technologies. If today the "romance" has gone out of industry, that is probably the greatest compliment we pay to the innovators of the past. We take for granted what they have bequeathed to us. A high level of material comfort is no longer lusted after; it is assumed.

Americans also assume that our material well-being will continue, will grow, and will diffuse to the rest of the world. But these assumptions will prove ill-founded unless a continuous supply of young talent is drawn to the invigorating challenges the chemical industry offers in such rich profusion. Hence the decision of SCI to produce this centennial record, in the hope that it may encourage creative young people of energy and imagination to take up the new demands being made on science, technology, and industry.

Today, the citizens of the world are growing in number at an unprecedented rate. And the expectation of long and healthy life is increasingly seen as a birthright. The chemical sciences and industries stand at the forefront of the search for ways to improve the quality of the environment and to find new weapons in the age-old battle against disease. A century from now the bicentennial of the American Section of the SCI will most likely recount how these challenges were met by creative chemists, engineers, and executives of the twenty-first century. We trust this history will help encourage new recruits to our noble cause.

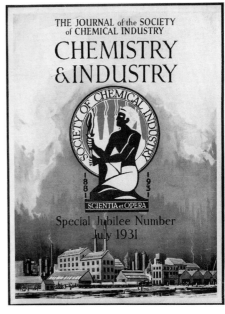

In 1931 the parent SCI celebrated its fiftieth anniversary with this special issue of the Journal of the Society of Chemical Industry.

HAROLD A. SORGENTI
Chairman, American Section
Society of Chemical Industry

5

Acknowledgements

Our thanks for making this publication possible go to John C. Haas and Ralph Landau, who created the nucleus fund for SCI's centennial celebration. Harold A. Sorgenti and J. Lawrence Wilson, immediate past chairman of the American Section, cheerfully raised additional funds that will make possible a fuller record of chemical innovation, through oral histories with Perkin and Chemical Industry Award recipients. We are also indebted for the editing and production of this history to Frances Coulborn Kohler assisted by Pat Wieland, and to Sylvia Barkan for its design.

ARNOLD THACKRAY
Executive Director
Chemical Heritage Foundation

SCI Project Supporters

The historical projects organized for the centennial of the American Section of the Society of Chemical Industry—which include the book *American Chemical Enterprise*—were made possible through the generosity of the following individuals, corporations, and foundations.

John C. Haas
Ralph Landau
Air Products & Chemicals, Inc.
American Cyanamid Company
Amoco Chemical Company
ARCO Chemical Company
Aristech Chemical Corporation
The Ashland Oil Foundation
The BFGoodrich Company
Bristol-Myers Squibb Company
Chemical Manufacturers Association
The Dow Chemical Company
Dow Corning Corporation
DuPont Company
Eastman Chemical Company
Exxon Chemical Company
Exxon Research and Engineering Company
Freedom Chemical Company
Great Lakes Chemical Corporation
Hoechst Celanese Corporation

Huntsman Chemical Corporation
IBM Corporation
The Lubrizol Corporation
Mallinckrodt Group, Inc.
Mobil Chemical Company
Monsanto Company
Morton International, Inc.
Nalco Chemical Company
Occidental Petroleum Corporation
Peridot Chemicals (New Jersey), Inc.
Phillips Petroleum Company
PPG Industries Foundation
Praxair, Inc.
The Procter & Gamble Company
Rohm and Haas Company
Shell Chemical Corporation
Solvay America, Inc.
Union Carbide Corporation
UOP
Chem Systems, Inc.

Introduction

*America created the modern technological nation. . . .
Inventors, industrial scientists, engineers, and system
builders have been the makers of modern America.*

— Thomas P. Hughes, *American Genesis* (1989)

The Parent Society Founded

The modern chemical industry and the Industrial Revolution were born to-gether in Great Britain at the end of the eighteenth century. Steam-powered machinery changed the world, and most immediately and dramatically it wrought a revolution in cotton textile production. That revolution would have been incomplete without cheap and effective chemicals for bleaching, dyeing, and cleaning fibers and fabrics. Responding to the challenge, British chemists and entrepreneurs pioneered in the large-scale manufacture of sulfuric acid, soda, chlorine bleach, and synthetic dyestuffs. During the nineteenth century, as the British chemical industry expanded, its leaders began to form organiza-tions to exchange views on business and technical matters and to enjoy the company of fellow gentlemen. Several such societies were founded in the 1860s, such as the Newcastle Chemical Society in the Tyne Valley and the Faraday Club, which drew its members from Widnes, St. Helen, and Liverpool. During an attempt to found yet another Lancashire society, a plan was broached to establish instead a nationwide organization to foster the exchange of ideas on topics related to the chemical industry. The broader plan came to fruition in 1881, when the Society of Chemical Industry (SCI) was inaugurated in the rooms of the Chemical Society in London.

The founders of the SCI were preeminent men in chemistry and industry. The first president was Henry E. Roscoe of Owens College, Manchester, the country's leading academic chemist. Other founding members included the en-trepreneur Ludwig Mond, who had introduced into Britain the superior Solvay process for making soda, and George E. Davis, the pioneer prophet for the as-yet-unknown, newfangled discipline of chemical engineering. Davis made sure that the SCI supported chemical engineering, even though a motion to call the group the "Society of Chemical Engineering" was defeated. Soon after this first meeting, the new SCI began plans for a journal that would keep its members

Henry Roscoe, the first president of the London-based Society of Chemical Industry, influenced generations of English industrial chemists who came under his tutelage at Owens College, Manchester. Courtesy of the Society of Chemical Industry.

One founder of the parent Society of Chemical Industry, George E. Davis, helped create the hybrid discipline of chemical engineering in the late nineteenth century, through his lectures to students in Manchester and through A Handbook of Chemical Engineering, *published in 1901. This "Double-Effect Vacuum Evaporator" is from the handbook.*

abreast of the latest developments in chemistry and chemical engineering. The first volume of *The Journal of the Society of Chemical Industry* was issued a year later, in 1882. Part of the journal's mission was to keep the members of the SCI's seven local sections informed about developments throughout Great Britain and the world. Not surprisingly, many aspiring American chemical entrepreneurs and small-scale manufacturers joined the SCI, if only to get its valuable journal. By 1894 there were 257 American members, over a quarter of whom lived in the New York area.

Early Days of the American Section

The organization of the New York Section—the name would be changed to "American Section" in 1919—of the SCI was instigated by Arthur McGeorge, an early Atlantic pond-hopping analytical chemist with offices in Liverpool and New York. He invited fellow New York members of the parent SCI to a meeting on 2 May 1894. Reflecting the chemist's role in contemporary American industry, the group that attended included two other consulting analytical chemists; one professor of analytical chemistry, who also did consulting work (Thomas B. Stillman, of Stevens Institute of Technology in Hoboken, New Jersey); three chemists associated with such characteristic local industries as creo-

Alfred H. Mason, first chairman of the New York Section of the Society of Chemical Industry. Courtesy American Institute of the History of Pharmacy Collection.

sote, printing ink, and varnish; and two British-born pharmacists. One of the pharmacists, Alfred H. Mason, emerged as the leader of the group. He worked for the Seabury and Johnson Company and in 1896 became the secretary of the New York College of Pharmacy. At a second meeting, with Mason acting as chairman and McGeorge as secretary, it was decided to send invitations to all the New York members of the parent SCI to attend a meeting at the College of Pharmacy headquarters on West 68th Street. There thirty-six members signed a petition to the Council of the SCI in London, stating that they wanted to found a section in New York.

Enthusiastically accepting the petition on behalf of the SCI Council, President E.C.C. Stanford welcomed the new section in ringing ceremonial prose: "We are pleased to add the stars and stripes to our highly respectable old colours, and to shake hands across the ocean with the great English speaking race who 'shall brothers be for a' that.'" Consonant with the purposes of the parent society, the U.S. section intended to bring together manufacturing and academic chemists for the exchange of information, but it did not want to challenge any existing organization. As the by-now-official chairman, Mason, stated:

> We believe there is room for our Society in America. It was founded, not for the benefit of a class, but rather for the purpose of blending together the manufacturing branch on the one hand, and the abstract on the other, and that amongst English speaking chemists. We have no wish to stand in the way of, or appear as rivals to, any other organisation, but rather as coadjutors with them.

The organization Mason wished to accommodate was probably the New York-based American Chemical Society (ACS), which required its members to have university degrees and exhibited an ambivalent attitude toward industrial chemistry. Mason expressed the wish that the new SCI section and the ACS could be mutually supporting:

> The American Chemical Society . . . has theoretical work to do, just as the Chemical Society in London has, and does it; and many members and fellows of both Societies are also members of this Society, recognizing that the industrial applications of chemistry have become so numerous that the existence of a separate body to especially consider this branch is desirable.

The rapid and dramatic increase in membership of the New York Section demonstrated the thirst for conviviality and practical knowledge among American chemical entrepreneurs. By the time that its petition was approved in London, the section's membership had risen to 310, which was more than 10 percent of the total membership of the SCI. When the first meeting of the new section was held in November 1894, 40 more members had joined.

Chairman Mason was clearly in an ebullient mood when he gave his inaugural

Ludwig Mond, another founder of the SCI, emigrated from Germany to England and established the firm Brunner, Mond, & Co. in 1873 to manufacture alkali by the Solvay process. Besides his role in founding the SCI, his legacies in his adopted country included the Davy-Faraday Laboratory of the Royal Institution.

address at that first gathering. He spoke on the lusty growth of the U.S. chemical industry reported in the 1890 census. He pointed to impressive increases in manufacture of chemical products since 1880. Fertilizer output had more than doubled, while sulfuric acid production had tripled. Mason was eagerly looking forward to the day that the power of Niagara Falls would be harnessed to provide cheap electricity for chemical production. His only discouraging report was that Americans had not made any headway with coal-tar products, especially dyes. The U.S. market continued to be dominated by German imports. In spite of this weakness and the ongoing industrial depression that had begun in 1893, Mason expressed confidence that American manufactures, including the chemical industry, would sooner or later achieve worldwide supremacy.

By the time Mason spoke, chemistry had become not only an established part of the curriculum of most U.S. colleges, but also a major segment of the still-small world of higher education. In 1890, 631 bachelor's degrees, 101 master's degrees, and 28 doctorates were awarded in chemistry. This represented nearly 10 percent of all undergraduate degrees, and 20 percent of all Ph.Ds., awarded that year. Other signs of the growing self-awareness and professionalism of chemistry and the chemical industry were the founding of the Manufacturing Chemists Association (1872) and the American Chemical Society (1876). These organizations had a strong presence in New York, which was the hub of the chemical industry in the United States.

In 1896 Charles F. Chandler, the next chairman of the new section, gave what was probably the best accounting for its success—the promotion of "friendly intercourse among the chemists of New York." An organizing genius, Chandler had been a founder of the ACS and its president in 1889, and he would become the first American president of the parent SCI. A Columbia professor for forty-seven years, he maintained such an extensive consulting practice that he was nicknamed "the godfather of the chemical industry." Of the New York Section of the SCI, he remarked:

Charles Chandler, the American Section's second chairman and the first American to be president of the entire Society of Chemical Industry, appears here with a washbottle. Chandler played many roles in the development of academic and industrial chemistry in the United States. Courtesy Chandler Museum, Columbia University.

The factory in which Bakelite, the plastic developed by the immigrant chemist Leo Baekeland, was produced. Chemical Heritage Foundation collection.

The Chemists' Club, the site of the meetings of the New York (later the American) Section of the SCI, moved to this impressive building in 1911. Left: the library. Below: the front door. Courtesy The Chemists' Club.

It serves to bring us together, to make us acquainted, and it enables us to help each other; and . . . for that reason, I think our Section has already been a great success, and I hope that it will continue to become more and more useful, not only to each of us individually, but to us collectively as a profession and to the community.

Chandler's words were indeed prophetic, if judged by the enthusiasm engendered for the New York Section. Members thronged to the monthly meetings from October to May to hear talks that were later reprinted in the parent society's journal. The list of early speakers included such notable foreign visitors as BASF's pioneer research director Carl Duisberg, who described the much-envied German system of educating chemists with the cooperation of industry. The attractions of America as a land of opportunity were symbolized by an address given by the young Leo Baekeland, recently arrived from his native Belgium. He spoke on his invention of Velox photographic paper, which had already earned him a fortune from George Eastman's Kodak Company. With that money Baekeland committed himself to a lifetime of chemical experimentation, the most important product of which was Bakelite plastic (see "Mauve and More"). Also making his debut in a new homeland was the biochemist Jokichi Takamine, who had been trained as a chemical engineer in Osaka and Glasgow before settling in the United States. Among his many accomplishments would be isolating adrenalin and engineering Japan's donation of cherry trees to the Jefferson Monument in Washington, D.C.

For chemists far from New York, who could not make the meetings, the journal of the parent SCI was an important resource. It contained articles not only from the New York sessions, but from all the British sections, abstracts of articles on industrial chemistry organized under twenty headings, statistical reports on the industry worldwide, and notices of new patents.

William Henry Perkin and his wife and two daughters on their American tour. Courtesy the Edgar Fahs Smith Collection.

William Ramsay, discoverer of the noble gases, Rudolph Messel, Anglo-German manufacturer of sulfuric acid, and William H. Nichols were eminent participants on the second occasion that the SCI held its annual meeting in New York, in 1912. Courtesy the Society of Chemical Industry.

The popularity of the New York Section's meetings in the 1890s exemplifies the city's growing financial, commercial, and industrial dominance in an industrializing United States. It was quite natural for Section members to take the lead in founding The Chemists' Club in Manhattan in 1898. Funds were gathered to rent the building of the Mendelssohn Glee Club on West 55th Street, where lectures and meetings could be held and a library of chemical books and journals maintained. The new Chemists' Club proved to be a hospitable location for meetings of the New York Section and the local ACS chapter as well. In 1911, when The Chemists' Club moved to its own building on East 41st Street, with beautiful facilities that included guest rooms and private laboratories, the New York Section continued to meet at the new site.

To complement the Section's fraternal activities, its chairmen occasionally suggested new projects, such as concerted lobbying on tariffs, patents, chemical education, and standardization of products. These suggestions generally fell on deaf ears, and the causes were left for other organizations to champion. Over time, even giving papers at meetings dwindled in favor of more informal proceedings. The American Section of the SCI thus found its niche in conviviality and communication, not lobbying or education. The core purposes of the organization—bringing together professionals for informal exchange of information, and promoting group identity as respite from business rivalries—were manifest in two major events in its early years. In 1904 the first annual meeting of the entire SCI to be held in the United States was convened. Two years later the New York Section helped orchestrate another Anglo-American event, a Perkin Jubilee to honor the venerable British chemist on the fiftieth anniversary of his epoch-making discovery of mauve.

The American Section Medals

The 1904 annual meeting allowed the Americans to advertise to their British counterparts the growing scale and sophistication of chemical manufacture in the United States. Approximately a hundred foreign guests were conducted on a one-month tour of nearly all the chemical manufacturing centers east of St. Louis, site of the extravagant Louisiana Purchase Centennial Exposition. At each stop local SCI members led field trips to such attractions as the Pacific Coast Borax Company and the Tide Water Oil Company, both in Bayonne, New Jersey; the new United States Mint in Philadelphia; the Bureau of Chemistry at the U.S. Department of Agriculture in Washington, D.C.; the Homestead Steel Works and the Nernst Lamp Company near Pittsburgh; the drainage canal in Chicago that kept sewage out of Lake Michigan; Parke, Davis and Company in Detroit; the Niagara Falls Hydraulic Power and Manufacturing Company in upstate New York; and the Harvard University laboratories, in Cambridge, Massachusetts.

At each venue the visitors were wined and dined and treated to light entertainment, including a special vaudeville in New York City that featured "The Journal of the Sassiety of Chomical Industry," issued by the irreverent New York publications committee. More serious business was the awarding of medals and honorary degrees. The parent society's medal was bestowed on Ira Remsen of Johns Hopkins University, the founder of graduate chemical education in the United States and the discoverer of saccharin. This incredibly sweet compound was actually manufactured not in the United States itself but in Germany, until the St. Louis-based Monsanto Chemical Company, founded by John F. Queeny in 1900, started production. At the end of the tour Columbia University presented honorary degrees to Sir William Ramsay, who would receive the Nobel Prize that fall for his discovery of the noble gases, and William H. Nichols, the second American to serve as SCI president, who headed a large sulfuric acid manufacturing company that introduced the modern contact process into America. Although Nichols had already enjoyed a prolific career in industry, his prominence would continue for another quarter century. In 1920 he engineered the mergers that created the Allied Chemical Corporation, for which he served as chairman of the board until his death in 1930.

The celebratory role of the New York Section was more durably established by the visit of the grand old man of the British chemical industry, William H. Perkin, in the fall of 1906. That year marked the jubilee of Perkin's discovery of mauve (aniline purple), the first synthetic dyestuff. In a bound volume presented at an earlier Perkin Jubilee celebration in England, the great German chemist Adolf von Bayer wrote that the aniline dyes are "the torch which enlightens the path of the explorer in the dark regions of the interior of the molecule," and that Perkin had lit that torch. Perkin's discovery also initiated a chemical industry based on research, as it pointed other chemists to the chemical riches that lay in coal tar. To mark the very special occasion of his visit, the New York Section decided to give the aged Perkin a medal.

The three medals struck by the American Section of the SCI. The Perkin Medal, established in 1906 (courtesy Research Corporation); the Grasselli Medal, established in 1920 (courtesy the Society of Chemical Industry); and its successor, the Chemical Industry Medal, first awarded in 1933 (courtesy H. L. Clark).

On 6 October 1906, at a banquet at Delmonico's in New York, four hundred chemical professionals celebrated the fiftieth anniversary of William Henry Perkin's discovery of the first coal-tar dye. In honor of the occasion, the guests all wore mauve bow ties. The bearded guest of honor, Perkin, is seated between Charles Chandler, the doyen of American industrial chemists, and a clergyman. To the clergyman's right sits William Nichols, a leading heavy chemicals manufacturer.

The idea of creating a "Perkin Medal" has been attributed to Herman A. Metz, who in 1906 was making a name for himself in the synthetic dye industry that Perkin had created fifty years earlier. The medal was to be awarded to "that chemist residing in the United States who had accomplished the most valuable work in applied chemistry during his career, whether this had proved successful at the time of execution or publication, or had subsequently become valuable in the development of the industry."

Although the Perkin Medal has always been considered a creation of the American Section, it was instituted by a joint committee that included representatives from the American Chemical Society and the recently founded Electrochemical Society as well. In subsequent years the Perkin Committee added representatives from the American Institute of Chemical Engineers, the American Institute of Chemists, and the American Section of the Société de Chimie Industrielle.

Upon arriving in New York, Perkin was feted as the guest of honor at a banquet for four hundred at Delmonico's, organized by the committee. There Perkin was presented with the first impression, in gold, of the medal. Two years later the first American award was made, to a longtime associate of Nichols's, J.B. Francis Herreshoff, who designed the equipment that brought the contact sulfuric acid process to America (see "Earth and Air"). The Perkin Medal would insure that the "chemical Edisons" of America would be duly recognized by a society that lionized inventors and innovators. Its significance was well caught by an editorial writer for the *Journal of Industrial and Engineering Chemistry:*

Davy cautioned Faraday not to indulge in immoderate expectations of the rewards of Science. . . . Science gives more precious rewards for excel-

Marston Bogert and the Mauve Tie

For twenty-three consecutive years, from 1930 on, the Perkin Medal was presented by a familiar and valued member of the chemical community—Marston Bogert, a former president of the parent SCI and an organic chemistry professor at Columbia. In 1953 the torch passed to another former SCI president, Wallace Cohoe, a chemical consultant. The 1953 Perkin Medal dinner was also notable for a new look; all the men were wearing mauve ties. Until that year the honor had been reserved for those present at the first ceremony, held for Perkin in 1906. By the 1953 award dinner, forty-six years later, only three men could legitimately wear the mauve tie: Bogert, Cohoe, and August Merz of Calco Chemical. The reappearance of a solid bank of mauve ties represented the passing of an era—and the beginning of another.

lence or success than those which can be expressed as ponderable quantities, and, among these few are valued more than medals. We look upon a medal as tangible proof of pre-eminence, for it is an indication of individual accomplishments—a token cast, in artistic mold, from a fusion of the golden opinions won by illustrious work. It furnishes the individual with evidence as to the regard in which achievements are held by his coworkers and fellow men.

To complement the highly esteemed Perkin Medal, in 1933 the American Section began to award the Chemical Industry Medal, to "a person making a valuable application of chemical research to industry. Primary consideration shall be given to applications in the public interest." In 1945, the criterion would be changed to "a person who . . . has rendered conspicuous service to applied chemistry." This medal replaced the Grasselli Medal, which had been given annually since 1920 for the best paper presented to the section. (The name Grasselli, which had been associated with heavy chemical manufacture since 1839, was fading from the chemical industry scene, since the Grasselli family had sold their company to DuPont in 1928.)

The first Chemical Industry Medal was awarded to James G. Vail, chemical director and vice president of the hundred-year-old Philadelphia Quartz Company, for his contributions to the chemistry and industrial applications of sodium silicates. Vail also performed valuable public service through his publications, which included the book *Sodium Silicates in Industry*, and his extensive commitments to professional associations.

Over the years the roster of winners of the Perkin and Chemical Industry Medals reads like a who's who of the American chemical industry. The careers of these medalists highlight the human drama that has characterized the ongoing development of the industry during the twentieth century. As our narrative will seek to show, the history of the industry can be told through the lives of the medal winners.

Spark of Genius

This drawing of the upper works of Charles Hall's Pittsburgh Reduction Company (circa 1900), adjacent to the generating house of the Niagara Falls Power Company, illustrates how favorable the location was to industry. Courtesy Alcoa.

Cover of the Carborundum Company's prospectus for its new diamondlike product. Courtesy The Carborundum Company.

The United States pioneered in electrochemicals, in part because it had abundant waterpower. It also had engineers who were expert at designing machinery. The States also produced young inventors, most notably Samuel F. B. Morse, Alexander Graham Bell, and Thomas A. Edison, whose imaginations were captured by the wondrous new phenomenon of electricity. Although the telegraph, telephone, and electric lights and motors were the most celebrated applications of electricity, chemical entrepreneurs were quick to utilize this unique form of energy to produce a wide range of products. Even before the dynamo made it possible to generate large quantities of electricity, batteries had been used to supply current for electroplating. When Niagara Falls was harnessed to generate electricity in 1895, it became a giant magnet for electrochemical enterprises. By 1902, sixteen firms lined the bluff above the gorge. The importance of electrochemicals in the United States is represented by its accounting for about twenty Perkin Medal winners in the first third of the century.

The first electrochemical pioneer honored with the Perkin Medal was Edward G. Acheson, in 1910. Raised in the coal fields of southwestern Pennsylvania, Acheson left school at age sixteen to help support his family after his fa

Edward G. Acheson in the lab with his omnipresent cigar, making tests on Aquadag, his soluble artificial graphite.

ther died. In 1880 he had the audacity to attempt to sell the great Edison an electrical invention; he instead landed a job at Menlo Park, helping perfect the new incandescent electric lighting system. After Acheson had learned the technology, Edison sent him to Europe to install electrical lighting apparatus, when he was just twenty-three. After he returned to the United States in 1884, Acheson left Edison to seek his own career in electrical inventing.

In 1891 one of Acheson's experiments yielded some small shiny specks of an extremely hard substance, silicon carbide, which he called Carborundum. Three years later he established the Carborundum Company in Monongahela City, Pennsylvania, to produce grinding wheels, whetstones, knife sharpeners, and other abrasives products. In 1895 he moved to Niagara Falls, where he produced other innovations, including artificial graphite.

In 1911, the year after it honored Acheson, the Perkin Medal committee chose Charles M. Hall for his 1886 discovery of an economical method for making aluminum. (The discovery was made simultaneously in France by Paul Héroult, who eventually met Hall and even spoke at his Perkin Medal ceremony.) The son of a Congregationalist minister, the young Hall became fascinated by chemistry when studying with Professor Frank F. Jewett at Oberlin College. After graduation Hall pursued the electrochemical production of aluminum, assisted by his sister Julia, who was also an Oberlin chemistry student. Although one of the commonest metals in the earth's crust, aluminum, "silver from clay," enjoyed the status of a semiprecious metal because it was so difficult to reduce to the metallic state. Its use was limited to such exotic purposes as capping the pyramid of the Washington Monument.

A contemporary described Edward Acheson's innovative characteristics:

It was imagination that led to his experiments; keen observation and appreciation of the value of the new thing that made him follow up the results he obtained; a knowledge of the art which directed the manufacturing processes he developed; and an extraordinary combination of optimism, self-reliance, determination, and perseverance which overcame the innumerable difficulties facing the creation of the new industry of artificial abrasives which have so profoundly modified a great number of manufacturing methods.

A more fitting description of the qualities of a Perkin Medal winner would be difficult to construct.

When Hall tried to create aluminum by electrolyzing an aluminum salt in water, all he got was aluminum hydroxide, because the aluminum ion reacted with the solvent water. He then focused on the crux of the problem: to find a solvent for aluminum oxide ore (alumina) that would not react with aluminum ions. Hall soon settled on cryolite—sometimes used as a source of aluminum—which melted easily. When he heated the cryolite in a crucible lined with carbon, the alumina dissolved; he then passed an electric current through it and aluminum formed: the experiment was a resounding success.

To scale up the process took years of work and a lot of money, however. In 1888 Hall teamed up with an experienced metallurgist with entrepreneurial yearnings, Alfred E. Hunt, to form the Pittsburgh Reduction Company. After exhausting their initial investment, the fledgling company was buoyed by resources from the Mellon banking interests. In 1907 the by-now-thriving enterprise was renamed the Aluminum Company of America (Alcoa). Hall, remembering his debt to Oberlin, made the college a major beneficiary of his estate.

Hall performed much of the research for Alcoa until his untimely death in 1914. The company president, Arthur Vining Davis, then sought to hire "an energetic, resourceful, chemical engineer" to set up and direct a corporate research laboratory. In 1918 he found Francis C. Frary, a thirty-three-year-old Ph.D. chemist from the University of Minnesota who had studied electrochemistry in Germany between his master's and his doctoral degrees. Under Frary's direction Alcoa researchers established a solid scientific understanding of aluminum that proved useful in virtually every aspect of the business. For his leadership and direction of Alcoa research for nearly thirty years, he was awarded the Perkin Medal for 1946.

Charles M. Hall in 1885, at the age of twenty-two, when he began to work in earnest on extracting aluminum from its oxide.

Charles Hall's sister Julia Brainerd Hall, who assisted him in his experiments, kept meticulous records of their work, and advised him on business negotiations.

Drawing of the original Hall electrolytic cell set-up in the Pittsburgh Reduction Company plant, showing the cast-iron crucibles or "pots"; the carbon anodes suspended by copper rods from an overhead copper bus; and, on the floor, ingot molds. On a good day the company produced fifty pounds of aluminum. All three courtesy Alcoa.

Rivaling the success of Alcoa is an enterprise begun by another pioneer electrochemist, Herbert H. Dow (Perkin Medal 1930). Dow, the son of an accomplished mechanic, grew up in Cleveland during the heyday of the oil and steel industries. Dow studied engineering at Case Institute of Technology, with the idea of emulating his father, but he became enthralled with chemistry. In 1888, while he was working on a senior thesis on boiler fuels, a driller gave him a bottle of brine with a substantial bromine content, typical of that found in many petroleum wells. Intrigued, Dow continued to investigate brines in Ohio and Michigan during 1889, while teaching chemistry and toxicology at a Cleveland hospital.

The relatively high bromine content of the brine underneath Midland, Michigan, attracted his attention. Local entrepreneurs were using scrap from the lumbering industry to boil the brine down until the sodium chloride salt crystallized, leaving a bromine-rich solution that was treated with oxidizing agents and distilled to yield bromine liquid. Dow realized that a large amount of energy was wasted in evaporating the brine and distilling the bromine liquor. He determined to accomplish these two tasks by electrolysis and an innovative air "blowing out" process. Although his early attempts failed, Dow moved to Midland in 1890 and perfected his process. He then achieved a product of pharmaceutical quality—for in the still-limited pharmacopoeia of the time, bromides were valued as sedatives.

Dow next applied his electrochemical techniques to sodium chloride to produce sodium hydroxide and chlorine for bleaching powder. Soon he diversi-

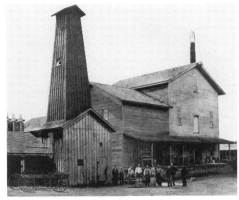

The old mill in which Herbert Dow set up his laboratory, which he rented when he came to Midland in 1890. Courtesy Post Street Archives.

Andrew W. Mellon, who bankrolled Hall, with his aluminum Pierce-Arrow before World War I. Courtesy the Family of Andrew Mellon. In general, U.S. automakers relied on steel instead, in part because they were more interested in low first cost than in low running cost.

Francis C. Frary, head of Alcoa's research and development, working at his rolltop aluminum desk during World War II. Courtesy Alcoa.

fied, first into chlorine chemicals, then into organic chemicals such as phenol, and finally to magnesium metal. He also laid the foundations for the long-term growth and success of his enterprise. His son Willard, who took over the company after Herbert Dow's death in 1930, received the Chemical Industry Medal in 1946, as did his son-in-law Leland I. Doan in 1964. Several other Dow chemists and managers have won SCI medals over the years including John J. Grebe (Chemical Industry Medal 1943), Edgar C. Britton (Perkin Medal 1956), Carl A. Gerstacker (Chemical Industry Medal 1974), Paul F. Oreffice (Chemical Industry Medal 1983), and Keith R. McKennon (Chemical Industry Medal 1994).

Another important chemical company, Union Carbide, also had its origins in electrochemistry. In 1892 at Spray, North Carolina, the inventor Thomas Willson attempted to produce metallic calcium—hoping that it would reduce aluminum oxide to the metal—by heating lime with tar in an electric arc furnace. Instead of calcium metal he made calcium carbide, which when dropped in water produced acetylene, already known as a remarkable gas that produced great heat and light when burned. Although his business enterprises were unsuccessful, Willson's process was critical to the new Union Carbide Company, formed from the merger of a number of small Niagara Falls companies in 1898.

A few years later Union Carbide acquired the Niagara Research Laboratories (NRL), a small venture company. One NRL founder, Frederick M. Becket (Perkin Medal 1924), would play a pioneering role establishing Union Carbide's research laboratories. Before becoming a research manager, Becket had an outstanding career as an innovator. Soon after he graduated from McGill University in 1895 with a degree in electrical engineering, he joined a New Jersey entrepreneur, Charles Acker, who had set up to produce aluminum

Herbert Dow's bromine plant at Midland, Michigan, in 1900. Left to right: the elevated electrolytic cells that oxidized the bromide dissolved in brine; the blowing-out tower that removed the free bromine from the brine; and the plant that extracted the bromine from the bromine-laden air, using iron or an alkali solution. To the right is one of the wells pumping up subterranean brine. Courtesy Post Street Archives.

by first producing sodium, much like Willson. When the success of the Hall electrochemical process became known, Becket went to Columbia University to study electrochemistry with Charles F. Chandler, while continuing to assist Acker and others. After Union Carbide acquired NRL in 1906, Becket perfected a process for reducing ores in a high-temperature furnace using silicon instead of carbon. This allowed him to make low-carbon steel alloys, such as ferrovanadium steel, used in automobile frames. He also developed techniques for reducing some of the rarer metals such as molybdenum, titanium, and chromium. In 1927, three years after winning his Perkin Award, Becket became the head of Union Carbide research laboratories, a position he held until his retirement in 1940.

Willson's calcium carbide process formed the basis for another important company—American Cyanamid. In 1898 the president of the British Association for the Advancement of Science, Sir William Crookes, warned that the world faced widespread starvation unless technology could be developed to "fix" the inert nitrogen in the atmosphere, making fertilizer for crops to feed the burgeoning population. One answer to the problem was the Haber-Bosch process for making ammonia from nitrogen and hydrogen (see "Earth and Air"). The cyanamide process, discovered by Adolf Frank and Nikodem Caro in Germany, was another. Frank and Caro patented their process, by which cal-

A corner of Dow's laboratory. Several important discoveries were made there by him and his employees, even though a certain creative messiness characterized research and development in these early days. Courtesy Post Street Archives.

A wagonload of bromides headed for Japan in 1908, with a new bromide plant in the background that still shows the characteristic blowing-out towers. Such shipments of bromides to foreign countries precipitated Herbert Dow's price-cutting "bromine war" with German manufacturers—which he won. Courtesy Post Street Archives.

My observation about the laboratory was that all the work was being done at the hood in one corner, where the chemists were in each other's way. Another hood or two would remedy this. The floor was poor and could not be kept clean, and is used as a highway. Covering with linoleum would correct this. The bleach sampler tramped in frequently with bleach on his shoes, which did not help the cleanliness of the place. This can easily be remedied by keeping him out. The table tops are so corroded as to make it impossible to keep them clean. They should be planed down, or, better, covered with glass or rubber. The latter would be comparatively inexpensive.— Albert W. Smith to Herbert H. Dow, 1910.

cium carbide reacts with nitrogen to produce calcium cyanamide, an effective fertilizer, in 1902.

In the United States a civil engineer and hydroelectric power executive, Frank S. Washburn, and a tobacco magnate, James B. Duke, teamed up to make cyanamide commercially in 1907. In 1912 they hired a Lehigh University metallurgy professor, Walter S. Landis (Chemical Industry Medal 1936; Perkin Medal 1939), to be their chief technologist. Landis was a Lehigh graduate and a protégé of Joseph W. Richards, one of the country's leading metallurgists and a founder of the Electrochemical Society. After joining American Cyanamid, Landis established a research laboratory and later served as a director and vice president for more than two decades. He received over fifty patents for his research, including several for processes that convert calcium cyanamide to sodium cyanide, used to increase the recovery of gold and silver from ores. His contributions to the fertilizer industry helped to prevent the Malthusian catastrophe that Crookes had predicted—and dramatically increased, with the Haber-Bosch process, the world's supply of fertilizer in the 1920s.

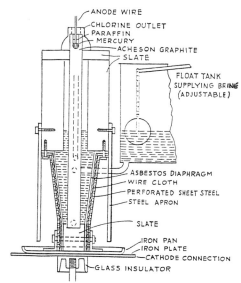

Designing efficient electrolytic cells engaged much inventive genius in the early years of the electrochemical revolution. These Moore-Allen unsubmerged diaphragm cells were among the achievements of Hugh K. Moore, who won the Perkin Medal in 1925. From Industrial and Engineering Chemistry *17 (1925) 207.*

Earth and Air

Winning Wealth from the Ground

Americans have long viewed their continent as an abundant source of wealth. From the very beginning of European settlement in the seventeenth century, the trees of the North American forests were harvested for lumber, fuel, and such chemicals as potash, methanol, and turpentine. The cleared land then became suitable for farming and grazing; and a cornucopia of mineral riches—from coal and oil in Pennsylvania to gold in California—lay beneath the soil. The country's young chemical industry naturally helped to make this underground wealth more accessible and valuable.

The search for minerals led two Perkin Medal winners, Auguste J. Rossi (1918) and John S. Teeple (1927), on adventures into the wilderness that symbolizes America's boundless riches. Rossi came to the United States from France, after earning his engineering degree. In the Adirondack Mountains of New York he encountered iron ore that was unusually rich in titanium. During the 1890s he succeeded in producing ferrotitanium alloys in a plant he set up at Niagara Falls—a feat that earned him the title of "father of alloy steels in America." During World War I, Rossi also pioneered in producing pure titanium dioxide, a superior white paint pigment that ultimately replaced the standard but poisonous white lead.

Instead of going north, Teeple, one of the outstanding consulting chemical engineers in the United States, went west into the desert. Teeple, who com

No Perkin Medal was given for innovations in the Solvay process until Robert Burns MacMullin was awarded one in 1971. The award was actually for a lifetime of process innovation, including work on the rival electrolytic technology, in which Americans had a greater role. In 1928 this young MIT graduate was asked by his employer, the Mathieson Company, to review the operations of its alkali plant in Saltville, Virginia, which had been erected on Solvay priciples in 1895. MacMullin's eyes bulged at the variety and complexity of the equipment required to carry out the dozen or so steps of the process. He streamlined the operation after performing a meticulous thermodynamic analysis.

In 1884 the Solvay Process Company opened this alkali plant near Syracuse on the Erie Canal and introduced Americans to their first large-scale continuous chemical process. Courtesy Mark W. De Lawyer.

The Trona plant of the American Potash and Chemical Corporation, designed by John Teeple. Courtesy Mark W. De Lawyer.

Teeple once commented:

If the [consultant] decides that, given time, patience, and money, particularly money that is patient, it is very probable, in fact a good bet, that a successful business can be built, if a little horse sense is mixed with the other ingredients. Then . . . you start picking men here and there who like work and responsibility. You show them your vision of what can be done. Soon there is a nucleus of men who have the vision and who attract other ambitious souls. The only bait needed is of a big pioneer work to be done, the promise that a man can have all the work, responsibility, and freedom from bossing and interference that he is capable of taking, and the assurance that the work will be completed. It is an adventure. Then you watch them grow, see them become an organization, a living, growing, cooperating entity working toward a definite end.

The adventure of innovation that Teeple so enjoyed inspired many other young chemists and engineers in the boisterous decades from the 1890s to the 1920s.

pleted his Ph.D. at Cornell in 1903, was awarded the Perkin Medal for work carried out after 1919 on the recovery of potassium chloride, borax, boric acid, and other chemicals from dry lake beds at Trona, California. He determined equilibrium data for complex mixtures of salts in water and designed continuous-process equipment for handling large volumes of mineral-rich brine—transforming a nearly bankrupt operation into a profitable concern. In the words of Williams Haynes, Teeple displayed in his work a "searching thoroughness, backed by a rare and acute sense of business realities. His influence upon his profession was enhanced by the clear pungent style of his many articles and speeches in which he set forth his stimulating, common-sense ideas."

The very first American awarded the Perkin Medal also capitalized on American resources. J. B. Francis Herreshoff received his medal in 1908 for a long innovative career in sulfuric acid production and copper refining and smelting. Although his family had been world-renowned boat builders in Rhode Island for generations, young Francis, unlike his two brothers, was captivated by chemistry. Combining his family heritage of engineering with the

A fleet of motor trucks owned by General Chemical Company—employer of J. B. Francis Herreshoff, one of the largest American chemical companies prior to World War I, and a forerunner of Allied Chemical Company. From The General Chemical Company after Twenty Years *(New York, 1919).*

Phosphate mining in Florida in 1909, primitive by today's standards, produced millions of tons of fertilizer after the rock was treated with sulfuric acid. Courtesy IMC Fertilizer Group, Inc.

knowledge he gained as an instructor in analytical chemistry at Brown University from 1869 to 1872, Herreshoff became a pioneering chemical engineer. After working as an analytical chemist for several years, he was hired by William H. Nichols, who owned a sulfuric acid plant.

Historically, sulfuric acid was the first industrial chemical, that is, the first chemical produced on a large scale because of its varied uses in early industrialization. The initial breakthrough in its production dated from 1746, when John Roebuck in England scaled up the prevailing "domestic" technology, based on glass bottles, by reacting sulfur dioxide with water in specially constructed lead-lined chambers. A critical improvement was his use of nitric oxide to catalyze the slow reaction of sulfur dioxide with oxygen to form sulfur trioxide, which then reacts with water. John Harrison of Philadelphia introduced the lead-chamber process to America in 1793, founding a firm that survived until 1917, when it was sold to DuPont. The process underwent constant incremental improvement, both in scale and in the concentration of acid produced, as demand increased throughout the nineteenth century. After the Civil War sulfuric acid was needed to make superphosphate fertilizers for the depleted soils of the South, which had the necessary phosphate deposits. Sulfuric acid also began to find extensive use in the refining of petroleum to make kerosene.

After joining Nichols, Herreshoff first committed himself to improving the performance of the existing plant by careful process innovation, sharply increasing the purity of the acid produced. But more dramatic challenges were in store. Nichols soon decided to use cheap Canadian pyrites as a source of sulfur. In 1896 Herreshoff developed a highly successful burner to convert the pyrites to sulfur dioxide, which became the world standard. The furnace was adopted for desulfurizing ores in copper refining as well. (Herreshoff had

Herreshoff's pyrites roaster, with its stacked hearths and removable rotating rabbles, used to stir the ore. From D. M. Levy, Modern Copper Smelting *(London, 1912).*

25

earlier worked out a more efficient smelting furnace for copper, in 1883.) Herreshoff's work with various ores emphasizes the way in which chemical engineering can lead to product diversification: with the new furnaces Nichols first became a major producer of copper, then invested in silver and gold recovery.

A related metal, platinum, was the catalyst in a new process for sulfuric acid production, the contact process, first developed in Germany. In the early 1900s Herreshoff took on the challenge of engineering a similiar process, since Nichols found the price for a license demanded by the original innovating firm, BASF (Badische Anilin und Soda Fabrik), much too high. Herreshoff's design was so successful that BASF quickly settled a patent infringement suit it had brought and ordered eight Herreshoff contact units before World War I.

Sulfur played a central role in the career of another early Perkin Medal winner (1912), Herman Frasch, the son of a prosperous German apothecary. In 1868 Frasch, instead of fulfilling his father's plans for a university education, followed the lead of many young men in search of adventure and fortune and emigrated to America at age seventeen. In Philadelphia he immediately found work as an assistant to a professor at the College of Pharmacy. He soon was swept up in the enthusiasm generated by the burgeoning Pennsylvania oil industry. In 1877 he sold the patent rights for an improved process for refining paraffin wax to John D. Rockefeller's Standard Oil Company and moved to Cleveland to continue his petroleum research. In 1886, after an unsuccessful venture in Canada, Frasch was retained by Standard Oil to work out a way to remove sulfur from the crude oil obtained from the Lima field in Ohio and Indiana; he applied processes he had begun to develop in Canada. Until then Lima's high-sulfur oil could be used only as cheap industrial fuel. Between 1886 and 1894 Frasch radically improved the desulfurization, enabling Lima crude to be refined into more valuable kerosene.

The grateful Rockefeller offered Frasch an executive position at Standard Oil, but, like many of his contemporaries, he preferred the life of an independent researcher. His next project was to develop a process to mine sulfur found during petroleum prospecting in Louisiana. In 1892 he succeeded by pumping large quantities of superheated water into the ground, which brought tons of liquid sulfur to the surface. The long-held Sicilian monopoly on sulfur was now broken. Within a decade the Frasch operation was daily pumping 12 million gallons of water heated to over 300 degrees Fahrenheit in 130 steam boilers, while sulfur spewed forth from a profusion of well derricks, each producing 500 tons of liquid sulfur that was then held in 150-by-250-foot bins. The United States was rapidly transformed from a country that imported over 90 percent of its sulfur to a major exporter of that chemical. One wonders if the young Frasch dreamed on his boat trip to the New World forty years earlier that his work would have such momentous consequences.

A look at America's mineral wealth would not be complete without some account of the uses of its iron ore. Of all the immense industrial enterprises in

Herman Frasch. From Journal of Industrial and Engineering Chemistry *4 (1912) 144.*

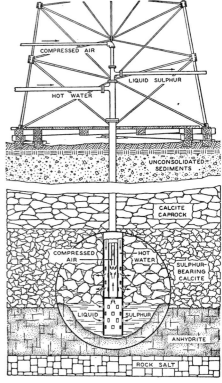

A sketch of the Frasch process for mining sulfur, which used superheated water to melt the sulfur and compressed air to force the molten sulfur to the surface. From Williams Haynes, The Stone that Burns *(1942). Copyright Texasgulf Inc.*

Left: One of Herman Frasch's sulfur wells in Louisiana, which discharged 500 tons per day of molten golden sulfur. The stacks of a battery of hot-water boilers are visible in background.

Below: A 100,000-ton block of sulfur. Both from Journal of Industrial and Engineering Chemistry 4 (1912) 145.

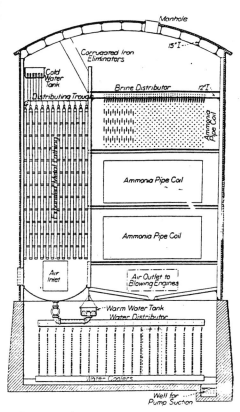

A diagram of James Gayley's improved two-stage refrigeration unit for drying out a veritable hurricane—the blast created by the blowing engine that heats the pig iron in a blast furnace. From Journal of Industrial and Engineering Chemistry 5 (1913) 244.

late-nineteenth-century United States, few could rival the size and importance of steel, the new building material used for everything from railroad rails to skyscrapers. Chemistry and chemical engineering proved essential to the growth of the country's steel industry. As early as 1876 the Pennsylvania Railroad hired Charles B. Dudley, a Yale Ph.D. in chemistry, to test, develop specifications for, and improve the materials used to build the railroads—steel rails and axles, as well as coal, lubricants, and the like. A decade later the steel magnate Andrew Carnegie hired James Gayley (Perkin Medal 1913) to manage the blast furnaces in his massive Edgar Thomson works near Pittsburgh.

The son of a Presbyterian minister from Lock Haven, Pennsylvania, Gayley earned a degree in mining engineering from Lafayette College in 1876. He then worked for a decade in the iron industry of the Lehigh Valley before heading west to Pittsburgh. While running Carnegie's blast furnaces, Gayley began to investigate a long-recognized problem: iron quality suffered during the humid summer months. Taking a scientific approach, he first made careful quantitative measurements of the amount of water that would have to be removed from the blast air to lower its humidity. It seemed technologically infeasible to solve the problem, which required removing more than 200 gallons of water per minute from a hurricane of air flowing at the rate of 40,000 cubic feet per minute. Gayley persisted in working on the problem, however, even after he was promoted into general management in 1896. By 1900 he was convinced that recently developed refrigeration technology based on ammonia could be adapted to drying the hurricane. Four years later he succeeded in building a plant on this principle, and the performance of blast furnaces everywhere improved dramatically.

Out of Thin Air

The use of ammonia for refrigeration, exploited by James Gayley, was part of the larger development of the industrial gas industry—a development in which Europe may have taken the lead, but one which soon made its way to this continent. Studies of gases and their role in chemical reactions initiated the chemical revolution of the eighteenth century that led to the establishment of modern chemistry. The industrial production of gases, however, did not begin until the late nineteenth century—with a few exceptions: "illuminating gas," made by controlled pyrolysis of coal; "water gas," made by passing steam over hot coke; and natural gas, at first consumed only near the wells that produced it. The large-scale use of other gases awaited the development of economical processes of production and safe methods of transportation—and sufficient demand.

Acetylene, for example, was first made commercially available after Thomas Willson's invention of calcium carbide (see "Spark of Genius"). It seemed ideal for use as an illuminating gas, but it was too unstable to store until the French engineer Georges Claude discovered that it could be put into cylinders safely when dissolved in acetone. Acetylene soon found an important new use in oxyacetylene torches, invented for cutting metals in France in 1901. A demonstration at the Brooklyn Navy Yard showed that these torches could cut portholes in three-inch armor plate in less than thirty minutes, a task that previously had taken five men nearly two weeks. The demand for oxygen—produced at the time by chemical means—also soared. In 1895 Carl von Linde, an engineering professor in Germany, had designed apparatus to liquefy air continuously through successive compression and expansion cycles. Once liquefied, the air could be distilled to separate oxygen from nitrogen. In 1902 Claude designed a similar process independently.

When inventors and owners of new European technologies were eager to export their processes to the United States, U.S. entrepreneurs were quick to establish the necessary companies. Gas technology was represented by the Linde Air Products Company, founded in Buffalo in 1907 and headed by Charles Brush. Brush also headed the National Carbon Company, which made graphite electrodes for electric arc lights and furnaces. Ten years later the Union Carbide and Carbon Company was formed by a merger of Linde Air, National Carbon, Union Carbide (calcium carbide; see "Spark of Genius"), Electromet (alloys), and Prest-O-Lite (acetylene). Linde Air Products (now Praxair) dominated the American industrial gas business for many decades.

In 1916 this experimental unit was constructed by the U.S. Department of Agriculture to study the then-revolutionary Haber-Bosch process of making ammonia directly, in a high-pressure catalyzed reaction of nitrogen and hydrogen.

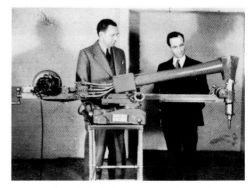

Air Products' Carl Anderson, chief engineer, and Leonard Pool, company founder, with their first order, for an oxygen cutting torch, in early 1941. Courtesy Air Products & Chemicals Inc.

A second American oxygen company, Air Reduction, was formed in 1915 by the French concern L'Air Liquide, which used Claude's technology.

In this early period most of the oxygen gas was used in oxyacetylene welding, and the nitrogen gas was used in the new General Electric lightbulb perfected by the chemist Irving Langmuir (see "Mortarboard and Lab Coat"). Another use was soon found for nitrogen in the Haber-Bosch process for making ammonia, developed in Germany by 1913; the ammonia was used to make fertilizers and explosives. This was the first high-pressure catalytic process, and it suggested a new approach to the production of many chemicals. It was also one technology not licensed in the United States, and the federal government and U.S. industry struggled to replicate it during World War I. In the 1920s the Allied Chemical Company and the DuPont Company built giant synthetic ammonia plants.

The Perkin Medal winner Frederick Cottrell—whose major contribution had been electrostatic precipitators to reduce smokestack emissions (see "In a Global Village")—accurately predicted in his acceptance speech in 1919 that some day oxygen would replace air in steelmaking and would be piped directly from large plants, not transported in cylinders. One restless and energetic entrepreneur who shared that vision and eventually made it come true was Leonard P. Pool (Chemical Industry Medal 1975). Pool, who had little technical training, started the Air Products Company in 1940 to make small on-site oxygen generators. His business flourished during World War II by making portable generators to supply pilots with oxygen for high-altitude flights. After the war Pool fulfilled his "piping" dream. Using new German technology to produce oxygen on a large scale, Air Products built enormous plants next to customers' sites and supplied oxygen via pipeline. In the 1950s Air Products diversified into liquid hydrogen, first for top-secret aircraft engines and later for NASA rockets. When Pool died, shortly after receiving his medal, his entrepreneurial vision had been realized in a large, diversified chemical company at the forefront of the by-then multibillion-dollar field of industrial gases.

Perhaps because so much of the early technology needed to produce oxygen and nitrogen originated in Europe, no Perkin awards were given in this area. Americans did excel, however, at producing noble gases, especially argon and helium. Floyd J. Metzger (Chemical Industry Medal 1934) was a pioneer in this area. After astounding Charles F. Chandler at Columbia by completing his Ph.D. degree in one year, in 1902, Metzger stayed on as a professor until 1917. He then joined the Air Reduction Company, where he developed the technology to separate argon, neon, krypton, and xenon from air. The commercial production of argon made it available as a replacement for nitrogen, which raised the efficiency of lightbulbs by 15 percent. Neon added a new element of color to signs everywhere. Metzger's work certainly made the world a brighter place.

Another rare gas, helium, was the focus of a crash program in World War I,

In the 1960s the fast food industry became a leading consumer of liquid nitrogen, used to freeze hamburgers, pizzas, and the like instantly. Here McDonald's beef patties are being frozen with liquid nitrogen. Courtesy Air Products & Chemicals Inc.

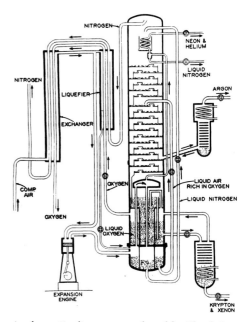

A schematic diagram employed by Floyd G. Metzger to explain how the atmospheric gases can be distilled from liquid air (1935). Courtesy BOC.

A Navy blimp filled with helium watched for submarines while escorting an Atlantic convoy during World War II. National Archives photograph 80-G-65729.

and its eventual success led to a Perkin Medal for Richard B. Moore in 1926. Although Moore was born in the United States in 1871, his family moved to England in 1878. He studied with Sir William Ramsay—discoverer of the noble gases—at University College, London, and was greatly influenced by him. After several academic jobs back in the United States, Moore joined the U.S. Bureau of Mines, where in World War I he became intimately involved in the effort to extract helium from natural gas for use in airships. After the war he established a cryogenic laboratory to continue work on helium production, which reached commercial scale in the 1920s. The substitution of helium for flammable hydrogen in airships made them much safer, as was underlined in 1937 by the dramatic crash and explosion of the German hydrogen-filled *Hindenburg*, while landing at Lakehurst, New Jersey. When war with Germany came four years later, American airships filled with helium (denied to the Germans even before war began) played a key role in spotting enemy submarines.

An air separation plant in Silicon Valley, California, supplying nitrogen to be used as the inert atmosphere in which computer chips must be produced. Courtesy Air Products & Chemicals Inc.

Mauve and More

The modern organic chemical industry began with Perkin in England in 1856. In subsequent decades it gained momentum as more dyes were discovered and the science of organic chemistry was firmly established by August Kekulé and others. Armed with new understanding about the structure of organic molecules, European chemists turned coal-tar derivatives into myriad products, including pharmaceuticals, before World War I. In spite of its robust coal and steel industries, which produced enormous quantities of coke, the U.S. chemical industry did not make much progress in unraveling the subtleties of the unfamiliar coal-tar chemistry until the war interrupted shipments from Europe. Chemists in the United States at first concentrated instead on less complex processes that used the abundant output of agriculture—corn, cotton, animal fats, and wood—as raw materials for industrial products. The SCI medal winners reflect both this early emphasis and the transition to equality and even leadership in the area of synthetic organics by the end of World War I.

The second Perkin Medal winner, Arno Behr (1909), developed important new processes for extracting starch, glucose, dextrose, and oil from corn during the 1880s. At the medal presentation one of Behr's associates described a plant under construction in Chicago that incorporated his designs:

> It is the climax of the development of the industry of corn products; a proud monument to American enterprise, American skill, American ingenuity, American perseverance. It is a monument to the keen foresight of the progressive management of the "Great American Industry of Corn Products," and is no less a monument to the pioneers of the industry.

A few years later, in 1917, the Perkin Medal was awarded to Ernst Twitchell, for a low-temperature catalytic process for splitting animal fat into glycerin and fatty acids. Twitchell grew up in Cincinnati, where an extensive pork-processing industry provided fat for soap and candle makers. After graduating from the University of Cincinnati in 1886, he joined the long-established Emery Candle Company. In 1892 he developed his revolutionary process, which was adopted worldwide. The process not only simplified soap production, but allowed soap and candles to be made from low-grade fats, sparing the higher-grade fats for use as food.

One of America's most eminent chemists, Arthur D. Little (Perkin Medal 1931), began his career by attempting to convert another ubiquitous natural product, wood, into paper, using the sulfite process recently developed in Europe. Unable to afford his senior year at MIT, Little left school in 1884 and

Ernst Twitchell, whose catalytic process known as "Twitchellizing," greatly simplified soap production. Early U.S. industrial chemists improved processes based on natural products. Courtesy Cincinnati Historical Society.

Rohm and Haas was a manufacturer of organic chemicals based on natural products from its inception in 1907. Its first product was Oropon, a leather-tanning compound made from enzymes and inorganic chemicals—seen here in barrels at their Bristol, Pennsylvania, location. Courtesy Rohm and Haas Company.

Scientific production control of processed rubber at B. F. Goodrich in the 1930s, inspired by the leadership of George Oenslager in research and development. Courtesy the BFGoodrich Company.

Oenslager's discovery of organic accelerators drastically reduced the time required to vulcanize rubber—from ninety minutes to six minutes for an inner tube. Courtesy the BFGoodrich Company.

went to work in a Rhode Island paper mill that used the new technology. He soon made improvements in the process and was sent to North Carolina to construct a mill. Upon his return to New England, Little teamed up with his replacement, Roger B. Griffin, to form a consulting firm. As the only qualified American experts on the new sulfite process, Little and Griffin consulted with all sixty paper mills in operation in the early 1890s. After Griffin was killed in a mill explosion in 1893, Little carried on alone. His enterprise, a major pioneer in the new field of business consulting, prospered. Little himself became one of the most ardent supporters of basic chemical research in the United States, and his backing helped transform MIT into a leading research institution (see "Mortarboard and Lab Coat").

Another natural material, rubber, became critically important to the United States once Henry Ford made the motorcar affordable. As the automobile industry grew dramatically after 1900, the poor quality of rubber tires made flats an unwelcome and often dangerous event on many car trips. The research of George Oenslager (Perkin Medal 1933) was critical to the search for stronger and longer-lasting tires. Oenslager completed his master's degree at Harvard University in 1896, under America's first Nobel Prize winner, Theodore W. Richards. In 1906 he arrived at the Diamond Rubber Company of Akron, Ohio. Rubber processing had not undergone a major technological innovation since 1839, when Charles Goodyear discovered the basic vulcanization process, using sulfur and white lead. The lead was later found to be a catalyst in the reaction between the rubber and sulfur. At Diamond Rubber Oenslager set out to find a better catalyst, and in 1906 he discovered that two coal-tar chemicals, aniline and thiocarbanilide, not only greatly accelerated the vulcanization reactions but also acted as a preservative in the processed rubber. A few years later he made a second important discovery—that the addition of carbon black to rubber improved its strength and toughness dramatically. These two innovations were largely responsible for increasing the life of tires tenfold, from 2,000 to 20,000 miles. In 1912 the Diamond Rubber Company was purchased by the B. F. Goodrich Company, where Oenslager worked until he retired in 1939.

American chemists began to play an increasing role in the modern organic version of alchemy, the successful conversion of cheap and ordinary materials—even waste products—into more esoteric and valuable ones. Perkin himself was an early practitioner of this art: he discovered mauve while attempting instead to synthesize quinine, a chemical leached from the bark of a tropical tree and used to combat malaria. His starting material was aniline, one of the earliest chemicals isolated from the tar that formed as a by-product of making coke for fuel and coal gas for urban gaslights—a by-product at first used only for paving roads.

The first American contribution to the new synthetic chemistry came from John Wesley Hyatt (Perkin Medal 1914), who developed the first modern plastic, celluloid, in 1870. Legend has it that Hyatt, then a journeyman printer in Albany, New York, learned of a $10,000 reward offered for a substitute ivory to use in billiard balls. After trying several composites, he began experimenting with nitrocellulose, or cellulose nitrate—a soluble form of cellulose made by reacting it with nitric acid—discovered in Europe about 1840. Hyatt found that a mixture of nitrocellulose and camphor could be molded with heat and pressure to form a hard solid. Unfortunately, his "celluloid" did not make good bil-

William Henry Perkin, founding father of the synthetic dye industry, held a skein dyed mauve for this portrait, painted by Arthur Cope in honor of the fiftieth anniversary of aniline dyes. Courtesy National Portrait Gallery.

John Hyatt and his brother established the Celluloid Manufacturing Company in Albany and in 1872 moved manufacture to Newark, New Jersey. Courtesy Hoechst Celanese Corporation.

33

Even DuPont—known for nylon (see "A Symphony of Synthetics")—entered the synthetic fibers field with rayon. Here a rayon plant is under construction in Buffalo, New York, in 1920. Courtesy Hagley Museum and Library.

liard balls: they were slightly explosive! Undaunted, he continued to work with celluloid and found ways to make it look like exotic natural materials, such as ivory, tortoiseshell, and amber. He then developed a successful business making fancy goods of all kinds—combs, brushes, mirrors, small boxes, clocks, toys, and shirt collars among them.

As better moldable plastics were invented, celluloid might well have been forgotten had it not found a unique niche in the photographic business being developed by George Eastman. In 1889 Eastman began to use celluloid for roll film in his revolutionary new camera, the Kodak, which carried the slogan "you push the button and we do the rest." The Kodak snapshot opened up the vast potential market for amateur photography. A few years later Eastman's film was used by Edison to make motion pictures, and celluloid became synonymous with this new American institution. Yet as early as 1910 a much less flammable cellulose polymer—cellulose acetate—began to replace celluloid for movie film. Developed in Europe by the brothers Camille and Henry Dreyfus, cellulose acetate came to America during World War I for use as a flame-retardant coating on the cloth skins of airplanes. After the war Henry Dreyfus continued to develop a synthetic fiber made from cellulose acetate. In 1924 the first "Celanese" plant opened in Cumberland, Maryland, signaling the beginning of the American Celanese Corporation, now part of Hoechst Celanese.

Celanese was not the first chemically engineered cellulose used in fabrics. Silk, the shiny, elastic fiber produced by a special variety of caterpillar fed on mulberry leaves—and one of the most exotic and popular materials in the late nineteenth century—was the model. Several processes for making artificial silk

An advertising card for early celluloid products. Gift of William Helfand.

Baekeland with his wife Céline and their children, Nina and George, on a family outing at Snug Rock, Yonkers, New York, about 1900. Like several other chemical industry pioneers, Baekeland had a love affair with the automobile—which turned out to be an engine of progress in chemicals manufacture in the United States (see "Cracking Nature's Secrets"). Courtesy Smithsonian Institution.

from cellulose were developed in the 1890s—among them one using cellulose nitrate, devised by Hilaire Bernigaud de Chardonnet in France—but the resulting fibers were harsh and extremely flammable. Fine cellulose fibers also found a use in lightbulb filaments. A more successful fiber, viscose rayon, emerged in 1892, when Charles Cross, Edward Bevan, and Clayton Beadle in England discovered the viscose process, in which cellulose is dissolved in an alkaline solution to form a viscous syrup that is then extruded through pinholes into an acid that regenerates the cellulose as filaments. The large American rayon industry began in 1910, when the British silk manufacturer Courtauld's founded the American Viscose Company at Marcus Hook, Pennsylvania.

As for synthetic plastics, the next step after Hyatt was taken by Leo Hendrik Baekeland (Perkin Medal 1916), who in 1907 invented Bakelite, a polymeric plastic made from phenol and formaldehyde. Baekeland completed his doctorate at the University of Ghent in his native Belgium, then taught for several years. He arrived in New York in 1889 at the age of twenty-six, after winning a traveling fellowship to continue his study of chemistry. During that visit to the United States, Columbia's Chandler convinced Baekeland not to return to Europe and recommended him for a position with a New York photographic supply house. A few years later, while working as an independent consultant,

The original Bakelizer, used by Baekeland and his coworkers in 1907–1910 to form Bakelite by reacting phenol and formaldehyde under pressure at high temperature. Donated to the Smithsonian by Union Carbide, the successor to the Bakelite Corporation, it was recently named the first National Historic Chemical Landmark. Courtesy Smithsonian Institution.

In 1916 the U-boat Deutschland *evaded the British blockade to bring both synthetic dye-stuffs and pharmaceuticals, which the U.S. chemical industry had been discouraged from producing, to the still-neutral United States. Courtesy United States Naval Institute.*

Baekeland invented an improved photographic paper, Velox, that could be developed using gaslight instead of sunlight. In 1898 he sold this invention to Eastman for a reputed $750,000. He now had the financial resources to carry out a lifetime of experimentation.

Working within the long-standing tradition of finding synthetic substitutes for costly natural materials, Baekeland looked for a replacement for shellac, made from the shells of oriental lac beetles. He began to investigate the re-actions of phenol and formaldehyde, which produce extremely hard resins. By carefully controlling the pressure and temperature of the reaction in his "Bakelizer," he found that he could produce a resin that when mixed with fill-ers produced a hard moldable plastic. Ironically, one of his first customers for Bakelite was Hyatt, who used it in billiard balls! Baekeland's new product soon found many other uses, especially in the rapidly growing automobile and radio industries. Upon his retirement in 1939, Baekeland sold his successful com-pany to the Union Carbide Corporation. By this time periodicals such as *For-tune* were enthusiastically proclaiming the coming of a "plastics age"—an age pioneered by Hyatt and Baekeland.

Before it could enter the plastics age, the United States had to establish a coal-tar chemicals industry to supply the intermediate chemicals, such as phe-nol. World War I provided the impetus, disrupting European exports of coal-tar chemicals and encouraging American chemical companies to enter the field. When the British blockaded Germany after 1914, the supply of German dyes and pharmaceuticals slowed to a trickle. American firms now had to manufac-ture not only these specific German products, but also the intermediates for many U.S. products that had depended on imported German chemicals. Fi-nally, the war created an enormous demand for weapons made from coal-tar

Baekeland's laboratory at his home in Yon-kers, New York, where he carried out his ini-tial experiments with mixtures of phenol and formaldehyde. Courtesy Smithsonian Insti-tution.

Billiard balls were at the forefront of early plastics technology. Hyatt's nitrocellulose-based ball (shown left) occasionally exploded on impact, causing surprised patrons of west-ern saloons to pull their guns! In 1912 the Hyatt-Burroughs Billiard Ball Company switched over to Bakelite balls (on right, with box). Courtesy Smithsonian Institution.

World War I prodded many companies into manufacturing organic chemicals, including Dow, which succeeded in producing the first synthetic indigo in the United States—seen here exiting Midland, Michigan. Courtesy Post Street Archives.

Although it was without enthusiasm that DuPont agreed to supply the Allied forces with smokeless powder during World War I, once committed, the company expanded with a will, relying heavily on women workers. Here the nightshift of "bloomer girls" takes a well-earned break in 1918. Courtesy Hagley Museum and Library.

chemicals, among them high explosives such as picric acid and TNT (trinitrotoluene).

The country's chemical industry responded to the new conditions with incredible energy and determination. Some of the solutions depended on traditional American strengths. World War I—a galvanizing experience for many a young chemist—was also the breeding ground for a pioneering generation of research managers in the chemical industry. Milton C. Whitaker (Perkin Medal 1923) left his chemical engineering chair at Columbia University to help the Industrial Alcohol Company, located near Baltimore, turn molasses into alcohol and acetic acid, which were in short supply. Whitaker was captivated by industry; ten years later, in 1927, he joined the American Cyanamid Company as vice president for research and built a widely respected research organization.

Lack of chemicals and the need to arm spurred development of another research organization. World War I was fought with smokeless powder, or gun cotton, a form of nitrocellulose made by nitrating cotton fibers. DuPont, the largest U.S. producer of dynamite, had worked with the American military for decades to perfect the new powder. Even before the war the company began to diversify into other nitrocellulose-based products such as paints and celluloid plastics. Producing explosives for the Allies—and later for the United States—dramatically increased the size of the DuPont organization, including its research staff, and brought large profits. Looking beyond high explosives (and needing an important dye intermediate to manufacture them), the company launched a major research and development initiative into dyestuffs manufacture.

Women workers at DuPont, assembling caps and fuses for munitions. Courtesy Hagley Museum and Library.

ACETPHENETIDIN (Phenacetin)

"If we don't sell you, we both lose money"

IRON BY HYDROGEN

VANILLIN

MONSANTO CHEMICAL WORKS
SAINT LOUIS

One of Monsanto's earliest advertisements (ca. 1910). It was for his expertise in manufacturing vanillin—synthetic vanilla—that John Queeny, Monsanto's founder, lured Gaston DuBois from Switzerland. Courtesy Monsanto Company.

The DuPont venture into dyestuffs turned out to be the most difficult technological challenge the company had ever undertaken. Because the Germans had been so secretive about dyestuffs manufacture, DuPont had to learn in a few years all the know-how that the Germans had accumulated over fifty years. The task of managing this monumental R&D effort after the war fell to Elmer K. Bolton (Perkin Medal 1945), then thirty years old. Bolton had earned a Ph.D. in organic chemistry from Harvard, then studied with Richard Willstätter in Germany before joining DuPont in 1915. Bolton systematized DuPont's approach to dyestuffs manufacture, placing emphasis on technical and economic efficiency. After a decade-long struggle, which included the difficult feat of adding former enemies—German dye chemists—to the staff, the DuPont dye business became profitable. In accomplishing this goal, Bolton built a powerful organic chemicals R&D capability that served DuPont well in the future.

The need for coal-tar chemicals in weaponry spurred yet other initiatives to supply the requisite raw materials. For example, chemical recovery stills were placed on thousands of coke ovens at steel mills and other locations across the country. Enough toluene was recovered, but not enough phenol to meet demand; Dow, Monsanto, and other companies therefore turned to making phenol from the more abundant benzene.

At the Monsanto Company the burden of war production fell upon Gaston DuBois (Perkin Medal 1944), a Swiss chemist. The founder of Monsanto, John

---

Perkin and Nobel Together—Again: A New Synthetic Organic

In 1982 Herbert C. Brown of Purdue University became the fourth person to win both the Perkin Medal and the Nobel Prize (1979). The other three to accomplish this tour de force of technology and science were Irving Langmuir (Perkin 1928; Nobel 1932), Glenn T. Seaborg (Perkin 1957; Nobel 1951), and Paul Flory (Perkin 1977; Nobel 1974). Early in his career Brown decided to work on organoboranes (then laboratory curiosities) because he was looking for a relatively fresh area in organic chemistry, which he considered a mature field—essentially settled. That judgment was premature, as he soon learned when he developed compounds that gave synthetic organic chemists powerful new tools for investigating the complexities of organic molecules—especially the role of steric effects—and synthesizing new molecules. Brown also worked out practical methods for commercial-scale preparation and manufacture of these compounds—thus his joint recognition by the Nobel and Perkin Committees.

Queeny, met DuBois when on a trip to Europe to investigate the manufacture of vanillin, a vanilla substitute made from coal-tar intermediates. In 1904 DuBois joined Monsanto in St. Louis. In the following decade he helped the company diversify from saccharin to vanillin and caffeine, to phenacetin for fever, chloral hydrate for sedatives, glycerophosphates for tonics, and coumarin for flavor enhancement. During the war he spearheaded the drive to integrate backwards into the production of phenol and phthalic anhydride. Following this intensive short course in chemical R&D, DuBois became production manager. Of the several positions he held at Monsanto, he was best remembered for his tenure as an enlightened vice president for R&D, during which the company became a widely diversified chemical manufacturer. In World War I alone Monsanto's sales increased from one million to ten million dollars per year.

The American chemical industry emerged from World War I with dramatically improved technical and financial resources that made the interwar period an innovative and successful one. Indeed it was in the field of synthetic organic chemistry that American chemical industry was to witness its single greatest triumph—the discovery of nylon by Wallace Carothers in 1934 and the creation of what over fifty years later remains a four-billion dollar plus business for DuPont, which still holds over twenty-five percent of the world's market (see "A Symphony of Synthetics").

A container for Monsanto's first product, saccharin (see Introduction), made for the Chinese market during the 1920s. Courtesy Monsanto Company.

Mortarboard and Lab Coat

Chemical Education and Industrial Research

A Perkin medalist at the end of the twentieth century typically holds a doctorate in chemistry from an American university or is a chemical engineer, often but not always holding a Ph.D. The medalist probably works in the research laboratories of a large corporation, or possibly a university. But as the lives of the Perkin medalists display, the contexts in which innovative scientists and engineers were nurtured has changed over time, in response to the development of educational institutions, the growth of large corporations, and the increasing sophistication of science and technology. Nevertheless, by World War I the main outlines of the modern setting for innovation were remarkably clear.

Training for Innovation

Americans have long considered themselves to be a practical people—problem solvers. In the eighteenth century Benjamin Franklin showed how science could be used to help solve technical problems. His legacy was sustained by the Franklin Institute, founded in Philadelphia in 1824 to teach mechanics how to apply science to their work. American colleges, with a few notable exceptions, concentrated instead on the education of clergymen, teachers, and gentlemen; the generation of new science was left to European universities for much of the nineteenth century. This country benefited from European training, however, when adventurous graduates emigrated to the United States seeking opportunities to profit from their specialized knowledge. Perkin medalists who immigrated before World War I included Arno Behr and Herman Frasch from Germany, Leo Baekeland from Belgium, Edward Weston from

Before World War I, a bachelor's degree was enough to allow medalists to become successful innovators. Among them were J. B. Francis Herreshoff, above, and Ernst Twitchell, whose process made the fortunes of the Emery Candle Company, shown below in 1923. Herreshoff courtesy The Chemists' Club; Emery Candle courtesy the Cincinnati Historical Society.

Great Britain, Auguste J. Rossi from France, and Gaston DuBois from Switzerland.* The members of this group had a better technical education in chemistry than their peers born in the United States.

Of the American-born Perkin medalists active before World War I, Edward G. Acheson and John W. Hyatt never attended college, although they quickly hired degreed chemists and engineers to staff their companies. Among those medalists whose formal study of chemistry ended with the bachelor's degree were J. B. Francis Herreshoff, Brown University; Charles M. Hall, Oberlin College; Ernst Twitchell, University of Cincinnati; Hugh K. Moore, Arthur D. Little, and Henry Howard, Massachusetts Institute of Technology; Richard B. Moore and Carl S. Miner, University of Chicago; Herbert H. Dow, Case Institute of Technology; and John V. N. Dorr, Rutgers University. These industrialists did not need advanced training to be significant innovators because the riches of the American continent were relatively untapped and the application of scientific principles to a wide spectrum of enterprises had just begun. Armed with basic chemical information, innovators could make revolutionary changes in industry—a process John S. Teeple (see "Earth and Air") called "chemicalization."

There were trained engineers among the early innovators, and even *chemical* engineers, although no degree was awarded in chemical engineering in the United States until 1891. James Gayley, Lafayette College, and Walter S. Landis, Lehigh College, were mining engineers; Frederick M. Becket, McGill University, and Charles F. Burgess, University of Wisconsin, were electrical engineers; Thomas Midgely and Frank J. Tone, both Cornell University, were mechanical engineers. The two chemical engineers in the Perkin cohort who were active before or during World War I were both MIT graduates—Warren K. Lewis, who went on to get a German doctorate in chemistry, and Robert Burns MacMullin.

Technical education in the United States began to expand dramatically after Congress passed the Morrill Act in 1862, which provided federal land grants for schools that taught agricultural and mechanical sciences. Courses in the basic sciences, including chemistry, formed the core of the curriculum in the new state universities that became familiar institutions by the 1890s, especially in the Midwest. Many students enjoyed their introductory chemistry so much that they became majors. By 1900 American colleges were producing about 1,300 bachelor's degrees in chemistry per annum. In earlier days, chemistry had been neglected or packaged without much success as a genteel discipline for the mind, but by the end of the nineteenth century it became clear that chemistry could provide jobs for new graduates, not only in teaching but in the developing chemical industry.

Inspired by professors like Charles F. Chandler (Perkin Medal 1920), who taught at Columbia University from 1864 to 1911, young chemists sought careers in a wide range of industries including sugar refining, petroleum, illuminating gases, electrochemistry, sanitation, and the analysis of water and minerals. Chandler was an early prophet of the academy not as an ivory tower but

Arthur D. Little, another early medalist, late in his career, visiting one of his laboratories in his "research palace" on Memorial Drive, Cambridge. Courtesy Arthur D. Little, Inc.

*Medal dates are omitted for passing references in this chapter, but may be looked up in the lists at the end of the book.

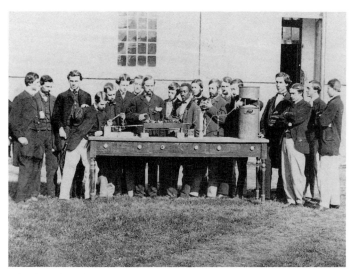

Charles Chandler and the Chemistry Society at Union College, circa 1863. Chandler held his first academic appointment at Union, where his position was budgeted under janitorial assistance. Courtesy Chandler Museum, Columbia University.

An assay laboratory, circa 1886, at the University of Pennsylvania. Chemistry instruction was often consigned to a lowly location— here the basement of College Hall—at colleges of the day. Courtesy E. F. Smith Memorial Collection.

as a place to learn how to apply knowledge to the real problems facing society.

Chemists who wished to pursue graduate work usually made the trip to Europe. Chandler himself earned his doctorate in the 1850s, with Friedrich Wöhler in Göttingen. Other Perkin Medal winners educated before World War I also studied in Germany, where Wilhelm Ostwald at Leipzig and Walther Nernst at Göttingen were gaining a more exact understanding of electrochemical processes and creating the new discipline of physical chemistry. Frederick G. Cottrell, Willis R. Whitney, Eugene C. Sullivan, and Colin G. Fink were Ostwald students. Irving Langmuir studied with Nernst, as did Sullivan before proceeding to Leipzig. The interest of these students in physical chemistry reflected the relative strength of the electrochemical industry in the United States.

One industrial weakness in the States was likewise reflected in the early Perkin Medals: few were given in organic chemistry. Americans had studied organic chemistry in Germany since the days of Justus Liebig, but they tended not to go into industry. One Perkin medalist, Warren K. Lewis, did earn a doctorate in organic chemistry, but it was as a preeminent engineer, not an organic chemist, that he rose to fame.

The obvious career choice for Americans returning from Germany with doctorate in hand was college teaching. As their number grew, the influence and energies of these academics helped establish graduate education in the United States—first at Johns Hopkins University, founded in 1876 on the model of the German university, with its obligation to perform research. In short order the doctorates began reproducing their like on home territory. By 1900 American universities were graduating about 70 doctorates and 150 master's degrees in chemistry per annum. Like their mentors, several of these newly minted American Ph.D.s then studied in Germany.

Relationships of professors with technology and industry were very much part of the German pattern. Nernst invented a lightbulb; Fritz Haber developed the ammonia synthesis; and Carl Linde built refrigeration and air liquefaction equipment. Of the Perkin medalists, Francis C. Frary studied the latest technology in electrochemistry at the Technische Hochschule in Berlin, and Elmer Bolton did postdoctoral work with Richard Willstätter on flower pigments. The profound interdependence of academic chemistry and industry that slowly developed in America is illustrated by the career of Ira Remsen at Hopkins. Remsen, the great pioneer in research as a tool of chemical education, was renowned for stressing the morally uplifting value of graduate education and the importance of "pure science" rather than practical application. Yet his own relations with industry increased as more and more of his students found employment there, among them the Perkin medalists William Burton at Standard Oil of Indiana and Charles M. Stine at DuPont. On retirement Remsen consulted for several firms, and in 1912–13 he served as president of the parent SCI, in London.

Chemical Engineering Invented

The most remarkable development in chemical education in the decades before World War I was chemical engineering, an American innovation. In England, George E. Davis had wanted the SCI to be named the "Society of Chemical Engineering." But the idea that he and others held—that a new kind of engineer was needed, one who understood both chemical processes and mechanical equipment—took firm root only in the United States. Implicit in Davis's teaching and writing was the notion that the number of distinct processes used in all the chemical industries—crushing, evaporation, distillation, and the like—was in fact limited. Traditionally, industrial chemists had been trained by learning the myriad manufacturing processes used in industries that ranged from leather tanning to food processing and copper smelting. Davis believed that these individual processes could be analyzed to discover general principles that could be applied in all chemical process industries. In 1916 Arthur D. Little (Perkin Medal 1931) coined the term "unit operations" to describe such processes.

The pioneer institutions offering chemical engineering were the Massachusetts Institute of Technology and the University of Pennsylvania. At MIT the famous Course X (ten) in chemical engineering was first offered in 1888, originally without its characteristic "unit operations" laboratory, and the first bachelor of science degrees in chemical engineering were granted in 1891. At Penn a bachelor's program in chemical engineering was begun in 1892.

At MIT, Little was particularly active in shaping the chemical engineering

Ira Remsen at the beginning of his long career at Johns Hopkins University, as chemistry professor and president. From 1876 on he offered the first doctoral program in chemistry in the United States, which he closely modeled on his own graduate training in Germany. Gift of Ned Heindel.

Students in a unit operations laboratory at Brooklyn Polytechnic Institute. Courtesy Polytechnic University.

The students in MIT's Course X in 1905. Warren K. Lewis is in the center of the back row. Introduced in 1888, the course marked the beginning of a chemical engineering curriculum at MIT. Courtesy the MIT Museum.

program around "unit operations" and industrial practice. In 1908, with William H. Walker, his second partner in his consulting firm, Little established MIT's Research Laboratory of Applied Chemistry as a bridge between academe and industry. In his scheme, companies supported fellowships for students who carried out laboratory investigations on that company's problems. In 1916 Little was instrumental in founding MIT's School of Chemical Engineering Practice, a program of industrial internships.

The beginning of the modern discipline of chemical engineering dates from 1910, when Warren K. "Doc" Lewis (Perkin Medal 1936) returned to MIT after completing his doctorate in organic chemistry at Breslau, then spending a year at a tannery in New Hampshire. Using the unit operations approach in his lectures, Lewis developed theoretical and mathematical concepts to explain unit operations on a more general level. In 1923 he and two colleagues wrote the first fundamental textbook on unit operations, *Principles of Chemical Engineering*. Complementing his command of his materials, Lewis's good-hearted banter with students created generations of loyal alumni, who had successful careers in academia and industry.

Employing Innovators: Chemists and Chemical Engineers in Industry

By 1910 an astonishing 16,000 chemists and chemical engineers worked in American industry. As early as the 1870s chemists acted as consultants to all manner of chemical process companies, giving advice on production problems or performing routine tests as the occasion arose. When a company needed

MIT's Warren K. Lewis teaching at the Chemical Engineering Practice School at the Bayway Refinery of Standard Oil of New Jersey. Courtesy Exxon Corporation.

In this barn behind Charles Steinmetz's house in Schenectady, Willis Whitney, a young MIT professor, began doing electrochemical research for GE two days a week during the winter of 1900. The barn burned down a year later, but GE research survived. Courtesy the GE Research and Development Center.

such advice frequently, it sometimes proved more economical to move the chemist in-house, as the Pennsylvania Railroad did with Charles B. Dudley—probably the first chemistry Ph.D. hired by industry—in 1876 (see "Earth and Air"). These early industrial chemists were generally not expected to develop new products, but to develop specifications and maintain quality, or to build plants and keep them running efficiently. Corporations usually relied upon independent inventors for major innovations.

Inventors could and did, like Perkin, form their own manufacturing companies. Among the early Perkin medalists were a number of such innovators: Acheson, Hall, Frasch, Hyatt, Baekeland, Dow, Weston, and Burgess are all noted above. Although less well known than some, Edward Weston (Perkin Medal 1915) and Charles F. Burgess (Perkin Medal 1932) were highly successful entrepreneurs. Weston made important improvements in electroplating before becoming a major rival of Thomas Edison in the electrical equipment and lighting business. Burgess left a professorship at the University of Wisconsin to work full time at invention. Among his many products was an improved dry cell battery. Although innovation was the very base of the companies these medalists founded, the notion that a company should be sustained by successive waves of innovation was not so deeply rooted. Like Edison, several of these inventors moved on to found other companies, propelled by their own creativity and impatience.

A few consulting firms founded before World War I also promoted innovation, along with other more traditional activities such as analysis. Among these were Arthur D. Little's company (see "Mauve and More") and Becket's Niagara Research Laboratory (see "Spark of Genius"). Two other innovative consulting firms with long histories were the Dorr Company, established by John Van Nostrand Dorr (Perkin Medal 1941) in 1910, which specialized in manipulating or treating particulate solids in liquids in continuous processes, and Miner Laboratories, which Carl Shelley Miner (Perkin Medal 1949) set up in 1906 to deal with agricultural products and processes.

Often university professors and their students served as consultants to industry, as in MIT's Research Laboratory of Applied Chemistry. Indeed, in the days before federal funding, most outside funding for academic research came from industry. A unique institution was the Research Corporation, founded in 1912 by Frederick G. Cottrell (Perkin Medal 1919), which received income from his highly lucrative patents for the electrostatic removal of particulate material from smokestack gases (see "In a Global Village"). The foundation's funds were then distributed to university researchers in the physical sciences as seed money—which bore fruit in many instances. Several other inventors contributed their patents to the Research Corporation, among them Robert R. Williams (see "Miracles from Molecules").

After 1900 the need for research within chemical industry became increasingly apparent, and a new institution was created, one that would prove an engine of innovation—the industrial research laboratory. These laboratories were initially modeled on the two precedents: Thomas Edison's "invention factory" at Menlo Park, and the research laboratories already established in German chemical, pharmaceutical, and electrical industries. The work of the new industrial research laboratories was intended to differ from the efforts of Edison and his assistants in being more continuously informed by scientific principles and in having improvement of product and process, not invention, as the goal.

Although by 1900 there were already many laboratories in the chemical process industries, they generally were part of the manufacturing division of a company. The first corporation to establish an autonomous research organization with its own director was General Electric (GE). In 1900 Willis R. Whitney (Perkin Medal 1921), then a chemistry professor at MIT, agreed to spend half of his time in Schenectady working on the critical problem then confronting GE. The original Edison lightbulb patents, which GE held, had expired, and numerous inventors were at work on new types of bulbs. Whitney's preparation for keeping GE abreast of the competition included an undergraduate degree in chemistry from MIT and his doctorate from Ostwald. At GE, however, Whitney learned that his talent lay less in original research than in the direction and inspiration of other chemists; he became America's first research director.

To conduct lightbulb research at GE, in 1907 Whitney hired Colin G. Fink (Perkin Medal 1934), who had just finished *his* doctorate with Ostwald. Fink worked with another chemist hired by Whitney, William D. Coolidge, on developing a ductile form of tungsten for lightbulb filaments. (After leaving GE in 1917, Fink worked for the Guggenheim interests on electrowinning of copper and then joined the faculty of Columbia University, where he served for thirty years.) The new tungsten filament was a great improvement over Edison's carbonized cotton filament; however, over time it became brittle and laid down a dark film of tungsten on the interior of the bulb. These problems were solved

Irving Langmuir and Willis Whitney with an early vacuum tube—the subject of several marketable innovations by Langmuir and also the source of many of his scientific advances. Courtesy the GE Research and Development Center.

by Whitney's most successful hire, Irving Langmuir (Perkin Medal 1928), who after his doctorate with Nernst had taken a teaching job at Stevens Institute of Technology.

To attack the problem, Langmuir concentrated on understanding the basic principles on which the lamp operated, investigating the chemical reactions catalyzed by the hot tungsten filament. He discovered that residual water in the bulb decomposed, and the free oxygen reacted with tungsten to create an oxide which in the evacuated lightbulb easily diffused to the surface and was deposited. Langmuir called this an "atomic ferry boat" and suggested filling the lightbulb with nitrogen to block the process. Since the nitrogen also caused convection that made the filament lose energy, Langmuir minimized it by reshaping the filament into the familiar spiral shape. This research created a superior lightbulb that generated large profits for GE in the following decades. The GE experiment in industrial research had paid off spectacularly, allowing Whitney to diversify his program to include new products. Langmuir in his turn continued his fundamental explorations of chemical reactions on metallic and other surfaces, for which he was awarded the Nobel Prize in 1932.

Other companies, large and small, were not far behind GE in realizing the potential of organized industrial research. Among the larger ones, DuPont, AT&T, and Kodak soon established their own laboratories. Smaller companies, such as Dow and Diamond Rubber, did likewise. The value of industrial research had become an article of faith among scientists, engineers, and businessmen in the United States. Cheered on by Arthur D. Little and other apostles of "the gospel of industrial research," U.S. companies set up about 300 industrial research laboratories before World War I.

For companies without in-house research facilities, the Mellon Institute outside Pittsburgh represented an attractive option. Andrew Mellon founded the

Rokeby Mill, temporary home of the DuPont Experimental Station, 1903–1906. Courtesy Hagley Museum and Library.

DuPont's Experimental Station in 1919. A new organic chemistry laboratory is partly obscured by trees in the middle of the photograph. Today the Experimental Station overflows the entire area captured in this photograph. Courtesy Hagley Museum and Library.

A view of Dow's Organic Research Laboratory in 1926, which was then ably led by William Hale and Edgar C. Britton (Perkin Medal 1956). It was described in company literature as "typical of Dow's modern research facilities." Courtesy Post Street Archives.

Mellon Institute of Industrial Research in 1913, as an institution that offered companies a team of Ph.D.s conducting research in well-equipped laboratories and experimental plant facilities—for a fee. (Its long-time director, Edward R. Weidlein, was awarded a Chemical Industry Medal in 1935.) The Mellon Institute possessed the collegiality of large university departments, but without students and with industrial research as its focus. George O. Curme (Perkin Medal 1935), recently returned from Berlin in 1914, noted that even while the Mellon Institute was housed in "little more than a wooden shack,"

> I was most impressed with the caliber of the group of thoroughly trained chemists who were conducting its affairs. I had wandered about quite a bit in my academic training, in graduate study at Harvard, the University of Chicago, and the University of Berlin, where the last word in prestige, personnel, and equipment had been the usual thing, but I had never before seen such determination and evident ability to do something worth while in industrial chemistry, and I liked it from the first day.

Curme remained at Mellon until 1920, when he joined the Union Carbide Company. There he worked on producing chemicals from petroleum, a field in which he became a pioneer (see "Cracking Nature's Secrets").

George O. Curme, Jr., and colleagues standing outside the laboratory, called "The Shack," at The Mellon Institute. Curme is second from the left; his brother, Henry R. Curme, wearing a laboratory coat, stands next to him. Courtesy Union Carbide Corporation.

Innovation Institutionalized

Reflecting the increased importance of the chemical industry in the United States, graduate chemical education flourished in the decades between the World Wars. The rapid advance of chemistry and the chemical industry continued to attract bright young people to the profession. The Morrill Act may explain why a disproportionate number of these chemists hailed from the Midwest, as students from farm families encountered chemistry at the new land

Throughout his long career at the University of Illinois, 1916–1957, Roger Adams built the chemistry and chemical engineering department into a veritable powerhouse of American chemical talent. Courtesy American Chemical Society.

grant colleges. Much of this chemical talent was then funneled into the University of Illinois, where Roger Adams (Perkin Medal 1954) and Carl "Speed" Marvel (Perkin Medal 1965) turned out large numbers of chemists and chemical engineers. Other schools, including Harvard University, Princeton University, University of Michigan, and University of California at Berkeley, as well as Columbia, MIT, and Wisconsin, also developed outstanding programs for graduate chemical education. American chemists no longer had to travel to Germany to obtain a first-class education.

In the 1920s several universities, such as MIT, Illinois, and Wisconsin, began to offer Ph.D.s in chemical engineering as well as in chemistry, as the discipline rapidly matured. Chemical engineering doctorates who won the Perkin Medal include Donald F. Othmer (Michigan), Ralph Landau (MIT), and John H. Sinfelt (Illinois). After World War I, Illinois and MIT granted the most doctorates in chemistry or chemical engineering to future Perkin Medal winners, with five each. Princeton is next with four. No other institution has produced more than two winners, indicating how widely diffused the spirit of innovation has become in American academe.

The roles played by Perkin medalists in industry have also changed. Large laboratories employing ever-larger research teams became the normal arena for innovation. This in part resulted from a movement to emphasize "fundamental research" that arose in the 1920s. The term went back to the rhetoric of Little and others who argued that practical problems could best be solved by the application of fundamental scientific principles. Proponents pointed to Langmuir's invaluable research at GE.

In 1926 the director of the chemical department at DuPont, Charles Stine (Perkin Medal 1940), convinced the company's executive committee that DuPont research also needed redirecting toward "pure science," especially in areas where academic science was not addressing problems of relevance to in-

Donald Othmer and Raymond Kirk helped define the discipline of chemical engineering with their authoritative Encyclopedia of Chemical Technology. *From* Chemical Engineering Progress *(Feb. 1988) 69. Used by permission of the American Institute of Chemical Engineers. © 1989 AIChE. All rights reserved.*

Charles Stine in his DuPont laboratory. Courtesy Hagley Museum and Library.

In 1936 Gaston DuBois, Monsanto's research director, was instrumental in acquiring the personnel and research facilities of the chemical consulting firm owned and operated by Charles Thomas and Carroll Hochwalt, here flanking DuBois. Gift of Carroll Hochwalt.

dustry. These fields included catalysis and polymerization. Stine told the corporate chieftains that the work of such a laboratory would contribute to the sum total of scientific knowledge and in the long run would allow DuPont to improve its products and processes.

Soon Stine had several groups of recent Ph.D.s working in a new laboratory that was informally referred to as "Purity Hall." One of these researchers, Wallace H. Carothers, helped establish the basis of modern polymer science and directed research that led to the discovery of neoprene synthetic rubber and nylon. After World War II, the word *nylon* itself came to symbolize the fantastic returns paid from investment in fundamental chemical research.

Starting up a manufacturing company based on an innovation became more unusual after World War I than in the early days of the chemical industry. One important exception among more recent Perkin medalists is Edwin Land, who founded Polaroid. Other medalists have maintained the old tradition of independent consulting and research companies. A prime example is the laboratory in Dayton, Ohio, owned by Charles A. Thomas (Perkin Medal 1953) and Carroll Hochwalt (Chemical Industry Medal 1971). Charles F. Kettering founded the laboratory. When Kettering established the General Motors Laboratory and moved to Detroit in 1926, Thomas and Hochwalt took over the Dayton lab and operated it independently until 1936, when they teamed up with Monsanto. The pattern of successful "Perkin" laboratories being acquired is not unusual. Milton Harris (Perkin Medal 1970) founded Harris Research Laboratories, ultimately purchased by the Gillette Company in 1955. The Houdry Process Company, started by Eugene Houdry (Perkin Medal 1959) collaborated with several oil companies, including Standard Oil of New Jersey and Sun Oil, then was bought up in 1961 by Air Products and Chemicals, Inc.

One research organization has a particularly unusual history. The Universal Oil Products Company (UOP) began as an independent organization and later was bought by a consortium of oil companies. In 1944 its stock was placed in a trust, with proceeds donated to the American Chemical Society to support fundamental research on petroleum chemistry—the origin of the ACS Petroleum Research Fund. Later the company became independent once again and eventually became part of the Allied Signal Corporation. Several important innovators have been associated with UOP, including the founders, Jesse and Carbon P. Dubbs, Vladimir Haensel (Perkin Medal 1967), and Edith Flanigen (Perkin Medal 1992).

After World War II chemical engineering design and construction firms became an increasingly important part of the industry. Scientific Design Company, founded by Ralph Landau (Chemical Industry Medal 1973; Perkin Medal 1981), constructed petrochemical plants everywhere. And the time-honored role of independent consultant has not vanished. Governments and companies from Israel to Burma and Japan sought the advice of Donald F. Othmer (Perkin Medal 1978)—on distillation operations of all kinds, the use of nonpetroleum fuels, desalination works, and many other topics.

A Payoff of Chemical Engineering: The Manhattan Project

World War II spurred several cooperative ventures between industry, academia, and the U.S. government. Three such wartime undertakings in chemical engineering—the push to produce aviation gasoline, the synthetic rubber project, and the penicillin project—are described in the next three chapters. Yet a fourth call upon the energies of industrial and university researchers and managers was one of the most difficult technological challenges ever undertaken, the building of atomic bombs.

Although it has been described primarily as a physics enterprise, the Manhattan Project was at least as much an achievement in chemistry and chemical engineering. Before 1940 it was thought that an atomic bomb could only be constructed with uranium 235, which somehow would have to be separated from the predominant isotope, uranium 238—a chemistry and chemical engineering problem of the highest difficulty. Heading the planning board that oversaw the various approaches was Eger V. Murphree (Perkin Medal 1950); other Perkin medalists on the board were Warren K. Lewis of MIT, with whom Murphree had studied, and George O. Curme.

One solution to the separation problem was gaseous diffusion, which relied on the very small weight difference between atoms of the two isotopes. A gaseous compound of uranium was created by reacting it with fluorine to form uranium hexafluoride. To investigate the feasibility of the gaseous diffusion process, the government asked the M. W. Kellogg Company to design and con-

Prince Bernhard of the Netherlands (right) is welcomed to the Rotterdam plant of Oxirane, a joint venture of Scientific Design Company and ARCO Chemical Company. Ralph Landau (second from left) is accompanied by the burgomaster of Rotterdam (left) and the plant manager. Courtesy Ralph Landau.

struct a large-scale facility (known as K-25) at Oak Ridge, Tennessee. At Kellogg the head of "process development" was Manson Benedict (Perkin Medal 1966), a physical chemist with a doctorate from MIT who had experience in petroleum research.

Also working on K-25 were Ralph Landau and Henry B. Hass (Perkin Medal 1968). Landau's responsibilities included arranging for production of the special chemicals—most notably perfluoroheptane and perfluoroxylene—needed to work safely with fluorine and the strongly fluorinating uranium hexafluoride. Hass, then a chemistry professor at Purdue University, directed research on this difficult perfluorination. Another medalist, Harold E. Thayer (Chemical Industry Medal 1976) was project manager at Mallinckrodt Chemical Company for the operations that provided the purified uranium oxide needed to produce the hexafluoride.

In 1945 the atomic bomb dropped on Hiroshima contained U-235 from Oak Ridge—produced by piggybacking the different engineering approaches tried out there. After the war Benedict briefly returned to the petroleum industry, then became a professor at MIT, where he organized the first nuclear engineering curriculum, and an important figure in the work of the Atomic Energy Commission. Hass left Purdue to work in industry and became director of chemical research at Kellogg, and Landau founded the Science Design Company.

The uranium oxide from Mallinckrodt was also needed to make enough uranium to create the atomic pile used in a second method of constructing an atomic bomb. This method had its origins at Berkeley in 1940, when Glenn Seaborg (Perkin Medal 1957), completing work begun by Edward M. McMillan, isolated a new fissionable element, plutonium. Plutonium does not occur naturally on earth except in minute traces, but is created when U-238 is bombarded

Glenn Seaborg in 1942, when he and co-workers at Berkeley isolated tiny quantities of plutonium, the first of the transuranium elements. In World War II plutonium was used to make the second generation of atomic bombs. Courtesy Lawrence Berkeley Laboratory, University of California.

When high-purity uranium in large quantities was needed for the Manhattan Project, Mallinckrodt supplied it, delivering the first sixty tons within ninety days of beginning the project. Courtesy Mallinckrodt Group, Inc.

with neutrons. For the Manhattan Project, Seaborg not only isolated the new element, but determined its structure and explored its chemistry. He was awarded the Nobel Prize in Chemistry in 1951 for his discovery of plutonium and other transuranium elements.

For the war effort, to produce enough plutonium to construct bombs required large reactors and remote-controlled processes for separating the element from the irradiated uranium used to produce it. When the government asked DuPont to take on this job, the company president, Walter Carpenter, assigned two of his top chemical engineers, Roger Williams (Perkin Medal 1955) and Crawford H. Greenewalt (Chemical Industry Medal 1952), who had just completed commercializing nylon. Even these experienced MIT-trained chemical engineers were at first daunted by the scale-up factor of ten billion required. DuPont nevertheless designed and built first a test facility near Oak Ridge and then the huge facility at Hanford, Washington, that produced the plutonium used for the first test at Alamogordo—and for the bomb dropped on Nagasaki that ended World War II.

After the war Williams and Greenewalt soon became members of the DuPont executive committee. Greenewalt later served as president of the company from 1948 to 1962. In the postwar era many chemical companies continue to play critical roles in the development of nuclear technology.

Cracking Nature's Secrets

In 1859 Edwin Drake discovered that large underground reservoirs of petroleum could be tapped by drilling into them. Within a few years entrepreneurs, including the young John D. Rockefeller, were distilling petroleum to produce kerosene for light and fuel. In subsequent decades Rockefeller gained ascendancy over this new industry by building large refineries and remorselessly extending his control over the chain of the operations that led from the oil fields to his customers. Having the luck of a Horatio Alger character, Rockefeller switched his business to a new product—gasoline—just when electric lights were eroding the demand for kerosene at the turn of the century. The age of the automobile arrived just in time for him.

From the very start, the automobile and the chemical industry have been partners in a synergistic relationship that has changed the nature of all our lives. As chemists and engineers have worked to improve the quantity and quality of gasoline, they have also discovered how to make chemicals in massive profusion from petroleum and natural gas. In its turn, the automobile industry has become a major motivator of the search for better fuels and catalysts. The impact of the automobile, cheap fuel, and petrochemicals on modern living standards has been profound. For example, air travel has become commonplace in part because of improved aviation fuel and lightweight plastics, both products of the economic and technological revolution based on oil and oil chemistry. The very fact that the United States was a laggard in the earlier phase of coal-based organic chemistry allowed American companies to take the lead in the petrochemical revolution that has shaped the half century since World War II.

One pioneer in this story was William M. Burton (Perkin Medal 1922), a student of Ira Remsen's at Johns Hopkins University (see "Mortarboard and Lab Coat"). Burton was the first chemist hired by Standard Oil of Indiana. In 1909 he and a team of researchers began working to increase the yield of gasoline from petroleum distillation. They succeeded by 1912, by developing a pressure-distillation process that "cracked" longer hydrocarbon molecules into the smaller ones that constitute gasoline. High-pressure technology was so new at this time that Burton adapted locomotive boiler designs for his stills. Over the next few years the father-and-son team of Jesse and Carbon Petroleum Dubbs improved upon the Burton process, achieving a continuous cracking process that separated the cracked fractions by weight. The Dubbses teamed up with the Chicago meat-packing magnate J. Ogden Armour to form the Universal Oil Products Company, which licensed the technology to the oil industry.

William M. Burton as a young man. Courtesy Amoco Corporation.

A Burton-Clark experimental cracking still, Wood River, Illinois. The still is constructed like a locomotive boiler to resist the high pressure used in cracking. Courtesy Amoco Corporation.

While Burton was improving the yield of gasoline from crude oil, George O. Curme (Perkin Medal 1935) was attempting to produce acetylene (see "Earth and Air") from petroleum, using a high-frequency electric arc. After finishing his doctorate at Chicago, Curme went to Germany to study with Fritz Haber and Emil Fischer, until the outbreak of World War I sent him home. He found a job with the Mellon Institute, where the Prest-O-Lite company had established a fellowship for research on new methods for making acetylene. The major problem with Curme's electric-arc process was that it produced significant quantities of ethylene, for which there were no uses. Curme then began looking into ways to produce other chemicals from ethylene. One observer, at least, of the post–World War II petrochemical revolution claimed Curme as a forefather: "He saw clearly in 1915 and 1916, before anybody else appreciated the possibilities, just what is happening today to aliphatic chemicals, and in bold strokes he outlined the giant synthetic organic chemical industry that would grow from his hydrocarbon work."

Curme's first breakthrough came during World War I, when he tried to make mustard gas from ethylene and instead synthesized ethylene glycol, eventually used in antifreeze (yet another chemical of obvious use to a growing automobile industry). Research shifted back to methods for producing acetylene and ethylene in 1917, when the Prest-O-Lite Company was absorbed during the formation of the Union Carbide and Carbon Company. Curme focused his efforts on cracking ethane to ethylene. Union Carbide also established a parallel program in the Linde Air Products laboratory in Tonawanda, New York, under the direction of James A. Rafferty (Chemical Industry Medal 1948). Encouraged by the progress of this research, in 1920 the company set up the Carbide and Carbon Chemicals Company and bought a gasoline refinery in Clendenin, West Virginia, to supply the necessary ethane and other light hydrocarbon gases—until then unused by-products of oil refining. Over the next decade the new company developed the ethylene glycol antifreeze, synthetic ethyl alcohol, and

Burton and Humphreys, pioneers of oil cracking, posed in 1946 in front of a photograph of a fluid-bed catalytic cracker—the technology that supplied most of the gasoline for motor vehicle use in World War II. Courtesy Amoco Corporation.

George Curme in his Mellon Institute laboratory before World War I, where he laid the foundation for the petrochemicals revolution that took place after World War II. Courtesy Union Carbide Corporation.

Dubbs process units at the Shell Company plant, Dominguez, California, in 1927. Courtesy UOP.

Encouraged by Curme's work at Mellon, in 1920 Union Carbide purchased a gasoline refinery in Clendenin, West Virginia, where Curme could conduct pilot plant operations on such products as ethylene glycol, ethylene dichloride, and isopropyl alcohol. Courtesy West Virginia State Archives.

The gas pump that mixed gasoline with additives, including tetraethyl lead and ethylene dibromide. The Ethyl Corporation, founded in 1924 by General Motors and Standard Oil of New Jersey, engaged major suppliers such as DuPont for tetraethyl lead and Dow for bromine. Courtesy Post Street Archives.

dozens of other industrial chemicals. The Union Carbide venture into aliphatic petrochemicals was a pacesetter for the whole chemical industry. Not surprisingly, it also produced four medal winners in Curme, Rafferty, and two individuals who started at the Clendenin works and later became Union Carbide executives—Ernest W. Reid (Chemical Industry Medal 1951) and Joseph G. Davidson (Chemical Industry Medal 1955).

For gasoline the next important improvement came in the form of an additive, tetraethyl lead, developed by Thomas Midgely (Perkin Medal 1937). Midgely had studied mechanical engineering at Cornell, until romantic interests swept him away to Dayton, Ohio, where marriage and a job with the National Cash Register Company awaited him. In 1916 he joined the Dayton Engineering Laboratories Company, formed by Charles Kettering, who had also worked at National Cash Register until he perfected his famous electric starter for automobiles. Working on a kerosene-powered light generator, Midgely encountered the phenomenon of engine knock, a phenomenon so violent that it could crack cylinder heads. For several years he studied knocking experimentally, searching for ways to eliminate it. By 1922 he and Kettering had settled on adding tetraethyl lead to the gasoline, which allowed a 25 percent increase in the horsepower and fuel efficiency of automotive engines, essentially eliminating the knock.

Neither Kettering nor Midgely stopped there: six years later they were looking for a safe substitute for the toxic and flammable gases (e.g., ammonia) then used in the still-fledgling fields of refrigeration and air conditioning. Midgely quickly determined that chlorofluorohydrocarbons—Freon—had the right combination of properties for safe and efficient cooling, and thereby set another industry on course. The Kettering laboratory in Dayton was a seedbed of innovation, and in the mid 1920s it was home to two future medal winners besides Midgely: his two young assistants, Charles A. Thomas (Perkin Medal 1953) and Carroll A. Hochwalt (Chemical Industry Medal 1971). Both Thomas and Hochwalt went on to play key roles in the success of the Monsanto Corporation after it acquired their laboratory in 1936.

The technical challenges of fueling the automobile continued to attract innovators. One such was the French émigré Eugene Houdry, whose development of catalytic cracking for gasoline earned him a Perkin Medal in 1959. In France, Houdry had taken a degree in mechanical engineering before going to work for the family metalworking firm in 1911. After World War I—in which he served in the tank corps and received honors for extraordinary heroism in battle—he pursued an interest in automobiles (especially racecars) and their engines. A postwar trip to the United States included stops at the Ford Motor Company and the Indianapolis 500 race. Houdry's interests soon narrowed to improved fuels. Because his native country had little petroleum—and the world supply of petroleum was thought to be nearing exhaustion—Houdry, like other chemists and engineers, sought to make gasoline from France's abundant lignite (brown coal). Houdry began working with silica-alumina catalysts to effect the hoped-for molecular rearrangement. In the late 1920s he shifted his feedstock from lignite to heavy liquid tars, and by 1930 he had produced small samples of gasoline that showed promise as a motor fuel.

Thomas Midgley and the engine in which tetraethyl lead was first tested in 1922. Courtesy American Chemical Society.

Houdry now looked to the United States. To build pilot plants, he teamed up in the early 1930s with two American oil companies, first Socony Vacuum and then Sun Oil. In 1937 the Sun Oil Company opened a full-scale Houdry unit at its refinery in Marcus Hook, Pennsylvania, to produce high-octane Nu-Blue Sunoco gasoline. By 1942 fourteen Houdry fixed-bed catalytic units were carrying the unanticipated burden of producing high-octane aviation gasoline for the armed forces. As American airplane production skyrocketed during World War II, to nearly one hundred thousand planes in 1944, demand for aviation fuel was met with the next generation of petroleum refining technology—fluid catalytic cracking.

This new process was in large part the work of Warren K. Lewis (Perkin Medal 1936) and Eger V. Murphree (Perkin Medal 1950). A serious limitation in the Houdry process was that it deposited coke on the catalyst. To remove the deposit required shutting down the unit and burning off the coke in a regeneration cycle. A more elegant version of the technique—the moving bed method—required two large vessels, one a reactor and the other a regenerator, with the catalyst circulating between them. But how to "move more than a boxcar of catalyst a minute," as Murphree put it, was a daunting technological challenge. Murphree had worked as a research assistant for Lewis at MIT in the early 1920s, then joined Standard Oil of New Jersey in 1930 to direct a group investigating the manufacture of chemicals from petroleum. He turned to Lewis for help on a new cracking process in the late 1930s—a collaborative effort that built on traditional ties between MIT and Standard Oil. Lewis and an MIT

Warren K. Lewis pointing out details of a catalytic cracking unit. Courtesy the MIT Museum.

colleague, E. R. Gilliland, came up with the idea of using a finely divided, "fluidized" catalyst, which could then flow to a regenerator when spent. A team of chemical engineers from across the petroleum industry worked out the plant design, and by 1940 a 100-barrel-per-day pilot plant was operating at Baton Rouge. Two years later the first full-scale plant went on stream. By the end of World War II the United States was producing ten times as much aviation gasoline with this new process as was forecast at the time of Pearl Harbor.

Another wartime expediency was to create extremely high octane aviation fuel by blending in substances made by catalytic polymerization and alkylation of basic petrochemicals. Monroe E. Spaght (Chemical Industry Medal 1966) made important contributions to this advance. Spaght earned a doctorate in physical chemistry from Stanford, then spent a year in Leipzig working with Peter Debye before joining Shell Oil in 1933. As head of the technical department for the company's West Coast refineries, he stepped into a leading role as the petroleum industry mobilized for World War II. Shell not only developed large-scale alkylation and polymerization processes but also produced specific chemicals such as cumene and xylidenes. As Spaght later recalled, "We found that we could make a chemical plant out of an oil refinery." After World War II, advances in petroleum-processing technology indeed made oil refineries into chemical plants. It was to build those plants that the MIT-trained chemical engineer Ralph Landau founded the Scientific Design Company in 1946 (see "Mortarboard and Lab Coat").

One important postwar breakthrough in petroleum technology was "Platforming," a catalytic "reforming" process (turning open-chain hydrocarbons

Workmen recharging the catalyst in a Houdry unit. Courtesy Sun Company.

At Wood River, Illinois, Shell's twin fluidized catalytic cracking units, shown in 1944, towered fifteen stories in the air. Courtesy Shell Chemical Corporation.

The billionth gallon of aviation fuel being drawn at the Marcus Hook refineries during World War II. Courtesy Sun Company.

During World War II many oil companies developed new technologies to fulfill the demand for high-octane aviation fuel. Under an army contract Phillips Petroleum tested fuels over central Alaska to forestall problems caused by cold weather. Courtesy Phillips Petroleum Company.

into rings) developed by Vladimir Haensel (Perkin Medal 1967). The new process was a dramatic improvement over the early catalytic cracking methods exploited in the Houdry and the fluidized-bed processes. Haensel first worked on reforming in 1935, in a summer job at the Universal Oil Products Company. He then took a master's degree at MIT and returned to Chicago in 1937 to work full time at UOP with Vladimir Ipatieff, a pioneer in the field of hydrocarbon catalysis—and to study for his doctorate with Ipatieff at Northwestern University.

Like Lewis and Murphree, Haensel attacked the problem of build-up of coke on the catalyst in reforming reactors. He discovered that if a platinum catalyst was used at high temperature with a high partial pressure of hydrogen in the reactor, no coke was laid down on the catalyst. Not only that, but the resulting gasoline had higher octane values than that produced by existing processes. After the platinum content of the catalyst was reduced to an economical level, Haensel's new process took the petrochemical industry by storm. Moreover, by converting the naphthenes in the feedstock to aromatics, Platforming could be used to make other petrochemicals such as benzene, which previously had been distilled from coal tar. In 1952 UOP and Dow combined efforts to develop the Udex process to extract aromatic chemicals from Platformed gas streams. Later, when a vice president at UOP, Haensel reflected on the source of his earlier success: "Contrary to popular belief, a good portion of research is not based on a few brilliant ideas, but on the recognition and utilization of many obscure leads."

Vladimir Haensel. Courtesy Vladimir Haensel.

Yet another medalist, John H. Sinfelt (Perkin Medal 1984), started research at Exxon in 1954 on improving the platinum catalysts that Haensel used in Platforming. Sinfelt believed that the overlapping chemistry and chemical engineering curricula at the University of Illinois, where he had earned his doctorate, had prepared him well for this assignment. After developing a new approach to bimetallic catalysts, which he called "clusters," Sinfelt invented a superior platinum-iridium catalyst for reforming that finally made it possible to produce high-octane gasoline without lead additives. His efforts ultimately meant that urban air pollution from automobile exhaust could be reduced significantly and fuel efficiency increased at the same time.

In the broader field of catalyst science and technology another line of research and innovation yielded important results. In 1985 the Perkin Medal was awarded to Paul B. Weisz for his career-long studies of natural and synthetic zeolite (hydrous silicate) catalysts. These catalysts have very high selectivity, facilitating only certain reactions between specific molecules. One use for them is in Mobil Oil's process for converting methanol to high-octane gasoline in one step. Another researcher, Edith M. Flanigen (Perkin Medal 1992), spent her career at Union Carbide and UOP, synthesizing intricate catalysts that opened up new avenues for catalytic reactions—not only zeolites but aluminophosphates and silicoaluminophosphates, which are less acidic than zeolites and offer limitless structures and surface environments. At Standard Oil of Ohio, James Idol (Perkin Medal 1979) invented a superior process for making acrylonitrile (used in manufacturing acrylic fibers) in 1957, only two years after taking his doctorate at Purdue. The new process catalytically synthesized acrylonitrile from propylene, ammonia, and air—instead of from more expensive chemicals—in a single step.

Finally, the career of James Roth (Perkin Medal 1988) demonstrates yet again the diversity of applications of catalytic reactions of hydrocarbons. After

Left: John H. Sinfelt in his laboratory at Exxon Research and Engineering Company. Courtesy John H. Sinfelt.

Right: Edith Flanigen and molecular models of an aluminophosphate and silicoaluminophosphate, new materials that show promise as catalysts. Flanigen is the only woman to receive the Perkin Medal. Photo by Tom Sobolik, Black Star. *Courtesy Union Carbide Corporation.*

A plant for making linear olefins for bio-degradable detergents, Chocolate Bayou, Texas (see "In a Global Village"). Courtesy Monsanto Company.

A Monsanto scientist holds a test tube containing bacteria that eat complex industrial waste. Courtesy Monsanto Company.

joining Monsanto in 1960, Roth used the technology pioneered by Haensel to solve an emerging environmental problem, the fouling of streams and lakes with detergents. Using a new platinum-based catalyst, he was able to convert branched-chain detergent molecules, which microorganisms found indigestible, to linear ones that were biodegradable. Commercial detergent manufacturers quickly switched to the biodegradable type.

Following this success, Monsanto began research on homogeneous catalysis, which involves soluble liquid-phase catalysts instead of the usual solid ones. Out of this work came a revolutionary process for making acetic acid by adding carbon monoxide to methanol in the presence of a catalyst. This process also quickly became the standard one. Using the same basic technology, Roth and his associate William Knowles turned to a very different application when they developed a rhodium catalyst to make the amino acid L-dopa, a drug used to treat Parkinson's disease. Their work was the first direct industrial production of an optically active molecule using a totally synthetic catalyst.

These recent elegant applications of catalytic technology show just how far petrochemical processing has come since Burton constructed his locomotive-shaped stills eighty years ago. The multiple feedback loop between the consumer's desire for more and better automobiles and the growth of petrochemical technology displays a synergy that has characterized the chemical industry ever since the first growth of textile manufacture and the concurrent emergence of such basic industrial chemicals as sulfuric acid and caustic soda. In the modern case of petrochemicals, research has led not only to superior gasoline, but also to a proliferation of polymeric materials that today serve many uses, from the high-tech to the mundane.

A Symphony of Synthetics

One of the great transformations in the chemical industry was the shift from making relatively simple basic chemicals to making the complex synthetic materials called polymers. This revolution—as represented by celluloid, rayon, and Bakelite—relied more on technological innovation than on scientific understanding of the materials, which was minimal. This situation began to change in the 1920s, however, when Hermann Staudinger introduced the idea that polymers were not, as previously thought, mysterious compounds held together by colloidal forces. He argued instead, to widespread ridicule, that they were merely long-chain versions of ordinary organic molecules. Staudinger's hypothesis was slowly becoming an accepted theory in the early 1930s, when DuPont's Wallace Carothers developed techniques for synthesizing long-chain molecules using standard organic chemical reactions. The emergence of polymer science in the 1930s coincided with a burst of innovation in polymer technology. That decade witnessed the discovery or commercialization—sometimes both—of a new world of polymer products, among them nylon, neoprene synthetic rubber, acrylic plastics, polyethylene, Teflon, PVC, polystyrene, and silicones. Incubated in industrial research laboratories rather than by lone inventors, and force-fed by the demands of World War II, these products were to symbolize the "better things for better living through chemistry" that defined the consumer-led society of the 1950s and 1960s.

The polymer revolution at DuPont was initiated by the company's central research director, Charles M. A. Stine (Perkin Medal 1940) who had hired Wallace H. Carothers and convinced him to do research on polymers. After Stine was promoted to the company's executive committee in 1930, he was replaced by Elmer K. Bolton (Chemical Industry Medal 1941; Perkin Medal 1945), the veteran research director of the Dyestuffs Department (see "Mauve and More"). Soon after assuming his new post, Bolton asked Carothers to investigate the chemistry of an acetylene polymer that he hoped might lead to cheap butadiene and perhaps to synthetic rubber. In April 1930 one of Carothers's assistants, Arnold M. Collins, isolated a new liquid compound, chloroprene, which spontaneously polymerized to produce a rubberlike solid. The close chemical similarity between the new substance and natural rubber encouraged Bolton to develop the polymer—neoprene—which became the first commercially successful specialty rubber, because it was superior to the natural product in some applications.

Two weeks after this discovery another Carothers associate, Julian W. Hill, produced a strong and elastic synthetic fiber during experiments designed to

A worker at the Bristol, Pennsylvania, plant of the Rohm and Haas Company holding a Plexiglas cockpit window, circa 1940. Plexiglas cast-acrylic sheet was the product of Otto Röhm's long investigations in acrylic chemistry. It was first manufactured in 1935 at Röhm und Haas, Darmstadt; production began at Rohm and Haas in Pennsylvania four months later. Courtesy Rohm and Haas Company.

produce superpolymers of high molecular weight. These early *polyesters*, however, had such low melting points and high water solubility that they had no commercial future. After a few attempts to solve these problems, Carothers dropped this line of research for several years. Bolton, however, encouraged him not to give up on the wider field of fibers. When Carothers finally renewed work in early 1934, he and his team soon discovered an outstanding polyamide fiber. Bolton played a key role in the development of the discovery, which would later be named *nylon*. In the next few years the periodic bouts of depression to which Carothers was subject worsened, crippling his scientific creativity and prompting his suicide in April 1937, just when the true magnitude of the discovery of nylon was becoming apparent. By this time Bolton had made the critical decision to commercialize nylon by aiming first at the lucrative silk stocking market, leaving other applications for later.

Bolton also decided that commercial nylon should be made from adipic acid and hexamethylene diamine, because these two exotic chemicals could be derived from benzene, the only widely available raw material with a six-carbon ring. His decision created innumerable headaches for two DuPont chemical engineers—Crawford H. Greenewalt (Chemical Industry Medal 1953) and Roger Williams (Perkin Medal 1955), both graduates of MIT. Greenewalt informally took charge. The obstacles appeared insurmountable: the two monomers were laboratory curiosities; polymerization technology was primitive; and the spinning of nylon filament proved problematic because the polymer began to decompose if heated to only a few degrees above its melting point. Although some of the early processes accomplished their goals only through what Greenewalt later termed "brute force and awkwardness," they worked well enough for the initial commercialization of nylon in 1940. This accomplishment also depended on the efforts of Roger Williams, who had spearheaded the com-

Elmer K. Bolton, DuPont chemical research director, 1930–1951, and winner of both SCI medals. Courtesy Hagley Museum and Library.

Nylon, discovered by Wallace Carothers and commercialized by Elmer K. Bolton, was soon commandeered for war purposes. Constructing a nylon parachute was an elaborate operation that required sewing together sixteen gores. Courtesy Hagley Museum and Library.

Operators in the polymerization unit of a DuPont neoprene plant in the 1940s control chemical reactions with the aid of recording instruments. Courtesy Hagley Museum and Library.

The pilot plant for "cold" synthetic rubber introduced by Phillips in 1944. "Cold" synthetic rubber had advantages over natural rubber and could be processed at much lower temperatures than earlier synthetics. Courtesy Phillips Petroleum Company.

pany's efforts in high-pressure technology in the early 1920s. Williams's long and difficult struggle to make ammonia and methanol, along with other high-pressure products, had generated a strong esprit de corps among his research staff. This spirit carried Williams and his staff through the years of heroic effort needed to perfect the processes for making the two nylon intermediates.

World War II accelerated the investigation of new synthetics. American polymer science and technology made great strides as polymers were called upon to replace metals in many uses, to perform unique functions in new technologies like radar, and to replace unavailable natural materials, notably rubber. The rubber problem was so critical to Allied fortunes after the fall of Malaysia that the U.S. government organized a large cooperative effort between the chemical and oil industries and academia. By the time severe shortages of natural rubber occurred in 1943 (as expected), synthetic rubber was being shipped in large quantities from new plants around the country.

The synthetic rubber research program pulled in both academic and industrial researchers, among them several future medal winners. Bradley Dewey (Chemical Industry Medal 1944) was a key administrator during the most difficult stages of the project. Since 1919 he had been president of the Dewey and Almy Chemical Company, which produced latex sealing compounds. Robert R. Williams (Perkin Medal 1947) ran the fundamental research section of the rubber enterprise. An expert on insulating materials, including rubber, Williams had been director of chemical research at Bell Laboratories since 1925. During

William Jeffers (second from left) and Bradley Dewey (right), the two directors who coordinated the synthetic rubber project, inspect distillation facilities. ©1943 Chemical and Metallurgical Engineering.

64

Phillips designed, built, and operated this plant for the government near Borger, Texas, to produce butadiene, a vital ingredient in synthetic rubber. The plant was completed in just sixteen months, in July 1943. Courtesy Phillips Petroleum Company.

the 1930s he privately pursued the structure and synthesis of vitamin B_1 (see "Miracles from Molecules"). Also involved in the project was Williams's colleague William O. Baker (Perkin Medal 1963), who had joined Bell Labs after finishing his doctorate at Princeton in 1938, at the age of twenty-three. Baker made many important contributions to polymer science and technology, then later became the head of Bell Labs.

Among the polymers deployed during the war effort were the silicones, which proved indispensable in aviation. Silicones—compounds first investigated in the nineteenth century—were taken up at the Corning Glass Company in the 1930s, under the prompting of Eugene C. Sullivan (Perkin Medal 1929), who had earlier developed Pyrex heat- and shock-resistant glass. From 1935 on, James Franklin Hyde (Perkin Medal 1971) led the research, manipulating silicon and carbon atoms to produce a variety of materials that could resist moisture and extreme heat and cold. Also working on the project were Rob Roy McGregor and Earl L. Warrick, on Corning Fellowships at the Mellon Institute. To manufacture Hyde's new polymers, Corning teamed up in 1940 with Dow Chemical, which produced the magnesium and the halides needed to make silicones. In 1944 the Dow Corning Corporation, of which Sullivan was president for many years, began industrial-scale manufacture of a silicone sealant needed for airplane ignition systems. At General Electric another chemist, Eugene Rochow (Perkin Medal 1962), had followed Hyde's work; he began to develop different types of silicones for use in insulation—filing five patents by 1940.

In the postwar economic surge, silicones and many other polymers would become big business: the long-heralded Age of Synthetics had finally arrived. Polymer science was also becoming firmly established in academia. The Perkin Medal was awarded to three academic chemists who helped build polymer science into a new discipline—Carl S. Marvel, Herman Mark, and Paul Flory.

During World War II William O. Baker, a polymer chemist at Bell Laboratories, contributed to the synthetic rubber research program by investigating the relationship of the two physical states of rubber, sol and gel. Baker discovered that the high temperatures used in factory drying of rubber, although they make the rubber more workable, produced too much gel, which is very rigid. The rigidity makes for unsatisfactory properties in the finished product. Gift of William O. Baker.

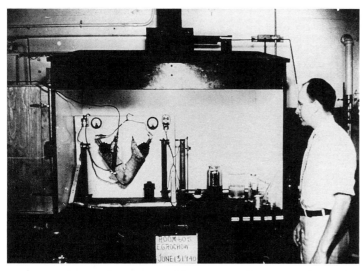

Eugene Rochow with a version of the improvised apparatus he used at the General Electric Research Laboratory in 1940. Rochow worked out the "direct process," a reaction that combines silicon with a compound of carbon, hydrogen, and chlorine to make silicones. Courtesy GE Research and Development Center.

James Franklin Hyde demonstrating a silicone-based product. Hyde began to develop silicones—products which became the foundation of Dow Corning Corporation—in 1935 at Corning Glass Company. Gift of James Franklin Hyde.

Marvel (Perkin Medal 1965) earned his doctorate in organic chemistry at the University of Ilinois in 1920, then stayed on to teach, becoming a professor in 1930. His introduction to polymers came in 1928, when DuPont asked him to investigate polysulfones. He then became a consummate synthesizer of polymers and educator of polymer chemists. His own technical contributions included solving several key problems in the synthetic rubber research project and, after the war, inventing heat-resistant polybenzimidazoles (PBIs), used in the space program. Marvel's most outstanding legacy, however, was his students—176 doctoral and 145 postdoctoral students. DuPont's excellence in polymer science and technology was in no small way due to the 46 Marvel Ph.D.s who joined the company's research ranks. His nearly sixty years of spreading the gospel of polymers made him a legendary figure in the history of the chemical industry.

Not even Marvel can quite challenge the record of Herman Mark (Perkin Medal 1980), whose name was virtually synonymous with the word *polymer*. An organic chemist educated at the University of Vienna, Mark in the 1920s studied the X-ray diffraction patterns of rubber, silk, and cellulose and confirmed Hermann Staudinger's theory that they were composed of high-molecular-weight molecules. He challenged Staudinger's view that they were rod-like structures, however, insisting that they were flexible. He conducted polymer research for the next decade, first at IG Farben and then back at the University of Vienna. After Austria joined Hitler's Germany in 1938, Mark was

Carl Marvel's career as a DuPont consultant, which began in 1928, was as legendary as his teaching career at the University of Illinois. Here he gives good advice to J. Burton Nichols at DuPont's Film Department in Buffalo, New York, in 1959. Gift of Carl S. Marvel.

Although Herman Mark agreed with Hermann Staudinger that polymers are true molecules, he disagreed with Staudinger's early rigid-rod model of them and held instead that they are flexible. Many years later in his office at Brooklyn Polytechnic Institute, Mark holds the two models that once were in question. Courtesy James J. Bohning.

dismissed from his post, emigrated, and took a job with a Canadian paper company. Two years later he found a permanent home at the Brooklyn Polytechnic Institute. For the next fifty years he was America's greatest promoter of polymer science and technology. In 1946 he established the Polymer Research Institute and the *Journal of Polymer Science*. By teaching, lecturing, and consulting, Mark shaped the direction of polymer research around the world.

A third academic, Paul Flory (Perkin Medal 1971) was the first American to receive a Nobel Prize for work in polymer chemistry; it was awarded in 1974 for his "fundamental achievements, both theoretical and experimental, in the physical chemistry of macromolecules." Like Marvel, Flory was initiated into the mysteries of polymers at DuPont, which he joined after completing his doctorate at Ohio State in 1934. Assigned to Carothers's group, Flory worked on problems related to nylon. Carothers also encouraged him to develop mathematical models of polymer chain formation; this led to his first published paper in a long and prolific career in academia and industrial research. He left DuPont for the University of Cincinnati not long after Carothers's death, then returned to industry (first at Standard Oil of New Jersey, then at Goodyear) during World War II, where he continued to work on a statistical-mechanical theory of polymer composition. After the war he joined the Cornell University faculty, then became director of the Mellon Institute, and finally moved to Stanford University. Among his contributions was *Principles of Polymer Chemistry*, a pioneering textbook and a classic in the field, which spurred the education of generations of polymer scientists and engineers.

As polymer science matured in the 1950s, synthetics arrived ever faster on the scene. Finer control over polymerization reactions, which theoretical work like Flory's promoted, meant that resins could be custom designed for specific properties. In 1954 Karl Ziegler in Germany discovered organometallic catalysts that created linear and stereoregular molecular chains. He and Giulio Natta shared the 1963 Nobel Prize for opening up this new field of polymer chemistry. The discovery also led to intense international competition as companies strove to reap the technological and commercial harvest from this break-

Paul Flory in 1955, operating the equipment he used to study the conformation of polymers at Cornell University. Courtesy Department of Manuscripts and University Archives, Cornell University Libraries.

Frederick J. Karol holding a model of a low-density poly-ethylene molecule. Courtesy Union Carbide Corporation.

Paul Hogan and Robert L. Banks, who in 1951 prepared crystalline polypropylene and then linear polyethylene. Courtesy Phillips Petroleum Company.

through. With Ziegler-Natta catalysts it became possible to make a truly synthetic rubber; a tougher form of polyethylene, without resorting to high pressure; and polypropylene, an entirely new polymer. This last discovery became the object of a thirty-year court battle between five companies. The patent was eventually awarded to Phillips Petroleum, which showed that two of its researchers, Robert L. Banks and J. Paul Hogan (joint Perkin Medal 1986), had actually produced polypropylene by another method in 1951. Banks and Hogan also developed commercial processes for making both polypropylene and the new linear high-density polyethylene.

Once it was shown that polyethylene could be made at much lower pressures than required in the traditional process, Union Carbide's Frederick J. Karol (Perkin Medal 1989) began to investigate making the original low-density polymer by similar methods. Karol joined Union Carbide in 1954 after finishing a bachelor's degree in chemistry at Boston University. Five years later he returned to formal studies, earning a doctorate in organic chemistry from MIT in 1962. The new polyethylene process occupied much of his time for the next fifteen years, until the first "Unipol" plant opened in 1977. The process was then licensed, and the production of the new polyethylene exceeded ten billion pounds per year by 1990.

The polymer revolution that began with Baekeland, Staudinger, and Carothers continues today, producing an ever-widening array of custom-designed chemicals to underpin the ongoing materials revolution of our day. No doubt polymer chemistry and technology will be well represented among the Perkin Medal winners of the coming century.

Miracles from Molecules

During World War I the American chemical industry scrambled to produce the new synthetic pharmaceuticals that were previously the monopoly of German companies. After the war the American pharmaceutical industry grew, but the United States continued to import more pharmaceuticals than it exported throughout the 1930s, a situation that changed after World War II. Among the SCI medalists are chemists who pioneered on several critical frontiers of modern medicine's struggle to alleviate the pain and suffering of injury and disease. Most Americans alive today recognize how indebted they are for health and longer life to anesthetics and painkillers, vitamins, antibiotics like penicillin or Terramycin, steroids like cortisone, and medications to reduce high blood pressure.

Over the years research and development in pharmaceuticals has become an extremely complex process involving large numbers of chemists, as well as biologists, pharmacologists, medical clinicians, and engineers. Chemists, with their ability to determine the exact composition and structure of molecules and to devise ways to synthesize them, have remained central to innovation in pharmaceuticals. Initially, chemists simply discovered organic molecules that "worked"—as in early anesthetics and sulfa drugs. They later imitated nature's own chemicals—and soon modified them in line with their growing knowledge of the relationship between the structure of a molecule and its chemical activity. More recently they began to design drugs that capitalize on the understanding of particular biochemical reaction sequences that occur in the body.

The chemical laboratory staff of G. Mallinckrodt & Co. in St. Louis, founded by the three Mallinckrodt brothers in 1897. Among its first products were aqua ammonia, spirits of nitrous ether, and acetic and carbolic acids to supply the pharmaceutical industry. Courtesy Mallinckrodt Group, Inc.

Merck's Rahway manufacturing site, circa 1903. In 1891 George Merck founded Merck & Co. as an entity independent from the firm E. Merck, Darmstadt. Like other American companies, Merck relied on German supplies of most organic chemicals until World War I, when this little plant was thrown into a frenzy of catch-up activity. Courtesy Merck & Co.

And not yet represented among the Perkin medalists are the biochemists and molecular biologists who are learning how to modify the workings of cells so as to produce particular products, such as insulin.

Anesthetics and Pain Relievers

The career trajectory of Ernest Volwiler (Chemical Industry Medal 1954) at Abbott Laboratories after World War I exemplifies the progression of the American pharmaceutical industry from imitation of German advances to native innovation. Volwiler was Roger Adams's first graduate student at the University of Illinois, completing his doctorate in the wartime year 1918. Adams was then consulting with Abbott Laboratories on three German drugs desperately needed on the battlefront: veronal, a sedative; novocaine, a local anesthetic; and atophan, a drug used to reduce fever and swelling. Adams convinced Volwiler to take a job with Abbott, where the rush to go into production with minimal pilot study was causing serious problems. Volwiler stayed in the works at night to supervise the reactions, snatching sleep on a cot. His efforts to compensate for inadequate equipment—and a workforce reduced by influenza—succeeded, and the products rolled out, greatly relieving the suffering of American battlefield casualties.

Abbott's management, notably Wallace C. Abbott himself and his successor Alfred S. Burdick, recognized the importance of research and development, and after the war they had Volwiler build up a research department. Volwiler and his colleagues used their skills in organic chemistry to perform hundreds of syntheses as they searched for new sedatives and anesthetics—of which only a few showed promise and were developed for market. One of the most

Stills producing ether at E. R. Squibb & Sons in 1912. Standardized pure ether was at the foundation of the company begun in 1858 by Dr. Edward R. Squibb, a former Navy surgeon. Courtesy Bristol-Myers Squibb.

The Abbott Alkaloidal Company was founded in Chicago in 1888 by Dr. Wallace Calvin Abbott to produce plant-based remedies in granule form instead of the more common fluid extracts, thus controlling dosages better. Left: a largely female workforce bottles and stores the granules, circa 1900. Below: Ernest Volwiler in the laboratory built for him in 1920. Courtesy Abbott Laboratories.

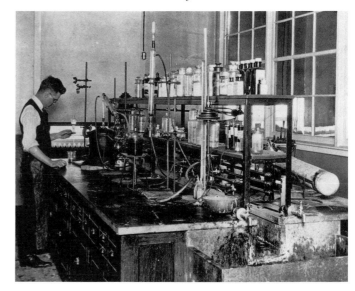

successful was Pentothal (sodium thiopental), an intravenous anesthetic introduced in 1936. As he rose in management—eventually to the presidency of Abbott—Volwiler retained his eye for a good new product.

Vitamins

Several medalists participated in the effort to ameliorate the vitamin deficiency diseases, whose causes were first identified early in the twentieth century. Surprisingly, the first such medal went to a chemical director at Bell Laboratories, not to an employee of a pharmaceutical company. Robert R. Williams was honored with the Perkin Medal in 1947 for his determination of the structure of vitamin B_1 and a synthesis that could be commercialized. He had lived his first ten years in India, where his father had been a missionary. After taking a master's degree in chemistry at the University of Chicago, Williams decided to return to Asia in 1908—this time to the Philippines. He managed to find a suitable post in Manila as a chemist at the Bureau of Science, a branch of the American colonial administration. There he was asked to determine the active principle in rice polishings that cures beriberi. He made a concentrate of rice polishings, which he himself administered to revive babies dying of beriberi, but could not then find the beriberi-preventing principle.

Through a long and illustrious career working for the U.S. Department of Agriculture, the Chemical Warfare Service, Western Electric, and then Bell Labs, Williams kept the problem of the beriberi-preventing principle in the back of his mind. In 1925, while still at Bell, he began to work on his old assignment in the evenings and on weekends in a garage in New Jersey, using his wife's washing machine as a centrifuge. Later Teacher's College at Columbia University gave him laboratory space. Eventually Williams turned to Merck & Co., whose managers had recently embraced the notion of basic research in pharmaceuticals. Merck agreed to supply the crystalline vitamin—which Williams had learned how to isolate by 1933—needed for further research; it also offered a newly equipped laboratory with qualified assistants. There in 1936 Williams made the structural determination of vitamin B_1 and designed a viable synthesis. During the course of his work he turned his patents over to the Research Corporation (see "Mortarboard and Lab Coat"), of which he later became head.

Merck began focusing on vitamins in the 1930s, when virtually all research on their structure and synthesis was conducted in Germany or Switzerland—though even there vitamins were not the ubiquitous dietary supplements they are today. Vitamin B_2 (riboflavin) was one target product, but IG Farben and Hoffmann–La Roche, the firms that held the patents for synthesizing it, would not license Merck to produce it. The job of designing a new synthesis of B_2 that would not infringe on the patents fell to Max Tishler (Chemical Industry Medal 1963), who in 1937 had just arrived from Harvard to be a research chemist. Like Ernest Volwiler twenty years earlier, Tishler personally supervised

Robert R. Williams and Robert E. Waterman (right). Courtesy Research Corporation.

The mystery posed by the high incidence of beriberi among people subsisting largely on white rice occupied Robert R. Williams for 25 years. Polishing rice to make it palatable removed some vital substance, which Williams and Waterman, Williams's son-in-law, confirmed was vitamin B_1. This 1937 photo above celebrates the successful synthesis of the life-saving substance. Under an agreement concluded in 1935, Williams and Waterman donated patent royalties on the synthesis of vitamin B_1 to combatting the diseases of malnutrition through Research Corporation's Williams-Waterman Fund. The fund was active for many years in Latin America and the Caribbean, contributing greatly to alleviating death and disability due to nutritional deficiencies. Over $10 million had been allocated by the fund when it concluded in 1978.

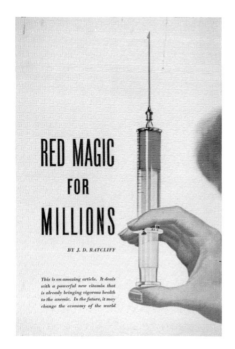

Left: Karl Folkers characterized the structure of vitamin B$_{12}$, advertised here, at Merck in 1948, and in 1949 the company began producing the red-colored vitamin as a treatment for pernicious anemia. Courtesy Merck & Co.

Right: Max Tishler, Merck chemist and research and development administrator from 1937 to 1969. Courtesy Merck & Co.

Below: As president of Abbott, Ernest Volwiler led the company into the post–World War II era of research, with its sophisticated technology such as this electron microscope—a far cry from his days at the laboratory bench. Courtesy Abbott Laboratories.

the scaling up of his synthesis to production scale—a job by then normally handed over to development specialists. But Tishler experienced the joy of being present when the first few drops of B$_2$ came out of the five-million-dollar plant. He was soon using his talents to head development, and later both research and development, for Merck.

The structure, synthesis, and commercial production of still other vitamins were rapidly mastered, at Merck and other drug companies. For example, in 1939, Karl Folkers (Perkin Medal 1960), a Wisconsin Ph.D., worked out the structure of vitamin B$_6$. A few years later he also worked on the cure for "pernicious anemia"—vitamin B$_{12}$. George R. Minot, working with Eli Lilly and Company, had determined that liver contains a factor that can cure this nearly always fatal condition, and in 1928 Lilly began extracting the factor from animal livers and selling it. Merck's Folkers characterized its structure in 1948, and Merck soon began producing it by a fermentation process.

Sulfa Drugs and Antibiotics

Despite strenuous efforts in Germany and elsewhere, creating chemical agents effective against the diseases caused by bacteria, viruses, and other microorganisms met with little success until 1932, when IG Farben patented the first of the sulfa drugs, discovered by Gerhard Domagk in their laboratories. Their drug, Prontosil, was effective against both streptococcal and staphylococcal infections. Soon pharmaceutical companies worldwide were making sulfanilamide, the product of the breakdown of Prontosil in human tissues, and other compounds containing the sulfonamide group. A number of SCI medalists were involved in these investigations. Max Tishler discovered a sulfa drug that was effective against a common chicken disease, coccidiosis. As directors of research, Ernest Volwiler at Abbott and Milton Whitaker (Perkin Medal 1923) at Lederle Laboratories encouraged investigation of the sulfa compounds in their laboratories, leading to general-use versions with fewer side effects and special-purpose forms for difficult diseases like leprosy.

In the 1940s sulfa drugs were joined by antibiotics, beginning with penicillin, discovered in 1928 by Alexander Fleming. Fleming observed the antibacterial action of a common mold, *Penicillium notatum*, that had accidentally grown in a petri dish prepared with a bacterial culture. But it was not until 1939 that Oxford's Howard Florey, with the assistance of his colleague Ernest Chain, extracted enough penicillin to allow clinical testing. In 1941 Florey traveled from war-torn England to the United States to enlist support in developing a large-scale manufacturing process.

In response to his presentations, a major cooperative program was established under the auspices of the American Committee on Medical Research and the British Medical Research Council. The penicillin program eventually involved hundreds of scientists, including several Perkin medalists, and some thirty-five institutions: university chemistry and physics departments, government agencies, research foundations, and pharmaceutical firms. Like the Rubber Program or the Manhattan Project, the wartime experience of the penicillin program stimulated both theoretical advances and production techniques exploited in future applications. Most of the chemistry departments and some industrial research groups focused on determining the structure of penicillin and finding methods for synthesizing it. While the Allied chemists succeeded in determining its molecular structure, the synthesis of penicillin proved elusive. It was manufactured by extraction from the natural mold grown in huge vats—

Below, left: Among those responsible for the production of wartime penicillin were A. N. Richards and Vannevar Bush, Office of Scientific Research and Development; George Merck, Merck & Co., Inc.; Sir Alexander Fleming; C. H. Palmer, E. R. Squibb & Sons; and John L. Smith, Chas. Pfizer & Co.

Right: The deep tanks used in submerged fermentation of Penicillium notatum *at Pfizer in the 1940s. This scaling up was achieved through modern bioprocess technology. Courtesy Pfizer Inc.*

itself an unprecedented technological accomplishment. Pharmaceutical companies like Pfizer, Merck, Squibb, and Abbott produced mountains of the little vials that saved the lives of countless men and women in the armed services—and, after the war, of countless civilians.

As good as it was, penicillin was not perfect. It had to be injected intramuscularly at frequent intervals, exhibited side effects in many patients, and showed little activity against some bacteria. As early as the 1940s researchers launched a great hunt to find other microorganisms that produced antibiotics. In 1943 Selman A. Waksman, a leading soil bacteriologist at Rutgers University, found streptomycin in an organism in the throat of a New Jersey chicken! Waksman's research was funded by Merck, so Tishler's people did the development work for this agent, which was effective against kidney infections, tuberculosis, and several other ailments that failed to respond to penicillin. At Waksman's request the patent was turned over to a Rutgers Foundation, and licenses were taken up by several drug companies, again in order to meet the demands of war and its aftermath of displaced and disease-ridden civilian populations.

After the war, pharmaceutical companies accelerated their search for antibiotics. Among the results of analyses of hundreds of thousands of soil samples from all over the world were Parke, Davis's Chloromycetin (1947), originating from a mold in a sample of soil from Venezuela; Lederle's Aureomycin (1948), from the dirt of an athletic field at the University of Missouri; Pfizer's Terramycin (1949), from the ground near the company's Terre Haute plant; and Lilly's Ilotycin (erythromycin, 1952), from the Philippines.

Peter Regna (Perkin Medal 1986) at Pfizer was among the chemists active in making structural determinations of Terramycin and other antibiotics. Studies of this sort formed the foundation for both improving the extraction or synthesis of natural products and creating synthetic analogues. After penicillin itself was finally synthesized in 1959, drug companies developed many penicillin-like medicines. Through the chemist's ability to change a cluster of atoms here or there, they could introduce drugs with greater effectiveness and fewer side effects than penicillin, such as penicillin V and ampicillin.

Steroids

In the 1930s organic chemists and biochemists recognized the structural similarity—the so-called steroid nucleus of four joined rings of atoms—of a large group of natural substances, including cholesterol, bile acids, sex hormones, the cortical hormones of the adrenal glands, and certain plant substances. In 1941 U.S. intelligence received reports that German airplane pilots were being fed an extract from ground-up bovine adrenal glands so that they could fly at greater heights. Although this information ultimately proved false, the U.S. government launched a program similar to the penicillin program, but on a smaller scale, to produce the hormone cortisone.

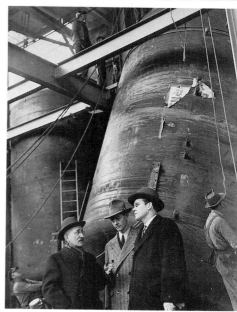

Selman A. Waksman (left) and Merck officials discuss progress on the installation of the 150,000-gallon fermentors used to produce streptomycin at Merck's Stonewall plant in Elkton, Virginia, in 1945. Above: Interior of a fermentation vat. Courtesy Merck & Co.

Robert Burns Woodward and two Pfizer chemists, Francis Hochstein and Karl Brunings, who led the team, which included Peter Regna, that determined the structure of Terramycin. In the early 1950s structural determinations—which are especially important for developing new medicines—still presented major difficulties. Terramycin's structure was not figured out until 1953, three years after it was first marketed. Courtesy Pfizer Inc.

Edward C. Kendall at the Mayo Clinic in Minnesota had already isolated cortisone and determined its structure. But there was no known way of producing it, except by the extravagantly inefficient method of extracting it from the adrenal cortex of animals. Merck, called in to find a better method, charged Lewis H. Sarett (Perkin Medal 1976), a new Ph.D. (1942) from Princeton, with the job. By the end of 1944 Sarett had created a thirty-seven-step synthesis of cortisone from deoxycholic acid derived from relatively abundant cattle bile. With the larger quantities now available, clinical testing began. By 1948 cortisone's role in alleviating the swelling of rheumatoid arthritis had been demonstrated, and clinical testing soon revealed many other uses. At Merck, Tishler was given the job of scaling up the synthesis for production—the most elaborate chemical process to be industrialized up to that time. Meanwhile, the search was on for a cheaper, shorter synthesis of cortisone—including a successful effort by Sarett himself, who ultimately led research and development for Merck.

Among Sarett's competitors was Carl Djerassi (Perkin Medal 1976), who was working at Laboratorios Syntex S.A., in Mexico City—a firm that the steroid chemist Russell Marker and a partner founded to produce synthetic hormones from steroidal substances in Mexican plants. Steroids had long fascinated Djerassi. Shortly after arriving in the United States in 1939, Djerassi sped through college, worked a year or two on antihistamines in New Jersey for the Swiss pharmaceutical company CIBA, and completed a doctorate in organic chemistry at the University of Wisconsin. He wrote his dissertation on how to transform the male sex hormone testosterone to the female sex hormone estradiol by a series of chemical reactions. When Djerassi returned to CIBA, he was not allowed to work on the synthesis of steroids because that promising topic was reserved for the laboratories at the corporate headquarters in Switzerland.

In 1944 Lewis Sarett holds the original sample of synthetic cortisone, which he and others learned to synthesize more quickly and cheaply by 1951. Courtesy Merck & Co.

Disappointed, in 1949 he joined Syntex, which was pursuing a synthesis of cortisone from diosgenin, a steroidal substance widely distributed in Mexican yams; Syntex had already produced male and female sex hormones from diosgenin. In 1951 Djerassi's group succeeded with a synthesis that not only started with a cheaper raw material, but involved about half as many steps as Sarett's original.

Soon other groups reported their syntheses. In the end, Upjohn was most successful commercially, with its use of a microorganism to convert progesterone to cortisone, but Syntex benefited, too, because it was commissioned to supply the progesterone to Upjohn. The next objective of the pharmaceutical industry was to reduce the incidence and severity of side effects suffered by patients taking cortisone. As with antibiotics, chemists soon learned how to synthesize other adrenocortical hormones and to modify their natural structures to improve their pharmaceutical characteristics.

In the same year that Djerassi's group synthesized cortisone, 1951, it also synthesized the first effective oral contraceptive. It was long known that during pregnancy progesterone maintains the proper uterine conditions and inhibits further ovulation, thus serving as a natural contraceptive. But taking natural progesterone orally weakens its biological activity. A more active sex hormone was needed that could survive digestive processes. One of the compounds synthesized at Syntex—in part for the Upjohn cortisone procedure—proved to be one of the most potent oral progestins ever made. In 1957 the Federal Drug Administration licensed norethindrone and a related drug that another company had produced, first as a treatment for menstrual difficulties and then as a birth control pill. Because Syntex did not at that time have production or marketing facilities, it distributed its product through Parke, Davis, among other companies.

A new cortisone synthesis is announced by Syntex, Mexico City, in 1951. Carl Djerassi is third from the right at the table, in the dark suit. On the table sits a Mexican yam, the source of the starting material for Syntex's cortisone and for norethindrone, the first effective oral contraceptive, also synthesized in 1951. Courtesy Carl Djerassi.

Djerassi himself, though maintaining a twenty-year-long relationship with Syntex, accepted academic appointments after his 1951 triumphs, first at Wayne State University in Detroit and then at Stanford University. He made many more advances in synthetic organic chemistry as well as refining instrumental methods for understanding the precise orientation in space of the atoms in a molecule. He also convinced Syntex to move its main research facilities to the Stanford Industrial Park to pursue new directions in pharmaceutical research, including applying molecular biology to drug design.

Designing a Drug

Understanding at the molecular level how living beings maintain the conditions of life has proved an excellent route to new drugs. One such is captopril, a blood pressure medication that Miguel A. Ondetti (Perkin Medal 1991) developed in 1966 at Squibb's research laboratories in New Jersey. Ondetti was educated as a chemist in Argentina and worked for Squibb there, then came to the United States in 1960 in the midst of a controversy about what causes high blood pressure and how best to reduce it and the threat it poses of heart attack or stroke. Of the three then-current theories about what causes hypertension—too much retention of water, too much force in the heart's pumping, and too much constriction of the smooth muscles of the blood vessel walls—Ondetti and the members of his group concentrated on the third. He wanted to find an orally active inhibitor of the action of angiotensin-converting enzyme (ACE), which converts the hormone angiotensin I to angiotensin II—the form that constricts blood vessels, degrades a hormone that relaxes the vessels, and stimulates release of a steroid that reduces the excretion of water and sodium from the kidneys. Ondetti found a naturally occurring ACE-inhibitor—a small molecule to occupy the enzyme's active site—in the bloodstream and then modified it so that it would bind more strongly to angiotensin I, creating a new approach to treating hypertension.

Ondetti's methods contrasted sharply with those of his predecessors, who depended for their initial discovery on accident or on mass screenings of compounds, synthetic or natural. His methods were closer to the work of those chemists who made substitutions for the side chains of this or that natural product; but even so, he and his assistants carried out precise reactions based on a much better knowledge of how the molecules of life actually interact. As knowledge of human biochemistry accumulates, the opportunities to engineer drugs that do exactly what is desired will become an increasingly important part of pharmaceutical research and development.

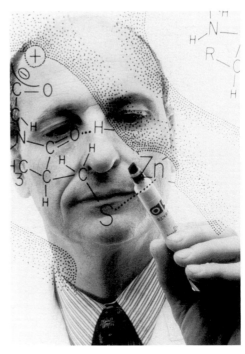

Miguel Ondetti focuses on a diagram drawn on plastic of the active site of angiotensin-converting enzyme (ACE) with a bound molecule of captopril, the specific inhibitor he and his colleagues designed as a new class of antihypertensive agents. Courtesy Bristol-Myers Squibb Company.

In a Global Village

We are indeed a prodigal people, prospering for a time by methods which would end European civilization within a generation.

—Arthur D. Little, "Making the Most of America,"
The Handwriting on the Wall (1928)

As we move toward the twenty-first century, the structure of science and technology has become truly global. The electronic information revolution—in which chemistry and chemical technology play essential roles—is knitting the world together as never before. Images and ideas travel everywhere and anywhere around the globe, in an instant. At the same time ecological research, much of it chemical, has shown that the earth operates as one large system—and that the scale of human activity has reached large enough proportions to affect that entire ecosystem. Carbon dioxide buildup in the atmosphere, for example, can lead to overall global warming, with potentially dramatic effects. The challenge for the future is to find ways to extend the globalizing power of the information revolution and maintain the benefits of modern life while ameliorating the detrimental effects of technology on the global environment.

The chemical industry has made many contributions to humanity's well-being—as glimpsed in the highlights traced in the foregoing chapters. Quite

A detailed shot of ENIAC's vacuum tubes. Courtesy Smithsonian Institution.

The mammoth size of early computers was largely due to their dependence on vacuum tubes—18,000 in ENIAC, the famous World War II computer. Transistors and later semiconductors changed the dimensions of computers radically. Courtesy of the University of Pennsylvania Archives.

naturally it has also created problems that demand solution. The solutions will come from chemists and chemical engineers, just as they have solved similar problems in the past—often by turning the unwanted by-products of chemical processes to good account as cheap raw materials. The careers of the Perkin and Chemical Industry medalists, stretching over the sweep of fourscore years and more, show that the capacity and the desire to solve environmental problems exists within the chemical community—perhaps uniquely so, through chemists' growing understanding of complex processes and new technologies.

A Revolution in Electronics

World War II accelerated the creation of the global village. In communications alone it not only spurred the development of a wide variety of sophisticated vacuum tubes, for uses ranging from radar to hand-held radio sets, but saw the beginning of research on the semiconductors that were to replace them in coming decades. Using the new science of quantum mechanics, physicists and chemists began in the 1920s to understand this unusual class of materials, which were neither conductors nor insulators. Silicon, which later gave its name to an industry in Silicon Valley, was the first to be exploited. In 1940 researchers at Bell Laboratories and MIT tried to make silicon rectifiers for radar, an idea suggested by the cat's-whisker rectifiers used in early crystal radio sets. They could not make the silicon pure enough, however, and it was chemists at DuPont, looking to make a white pigment, who succeeded that same year—making silicon with only one hundred atoms of impurity per million atoms.

The purity of the semiconductors used in transistors (and later in integrated circuit chips) was vital in the new electronics. Silicon of sufficient purity—even purer than the DuPont silicon used in World War II radar and far beyond the level of purity achieved by earlier electrochemists such as Frank J. Tone (Perkin Medal 1938) of Carborundum—was the achievement of N. Bruce Hannay (Perkin Medal 1983) at Bell Laboratories. During World War II Hannay worked on gaseous diffusion for the Manhattan Project, right after earning his doctorate in physical chemistry from Princeton. His first civilian project at Bell was investigating the mechanism of thermionic emission from the oxide cathodes of vacuum tubes, along lines similiar to those Langmuir pioneered at General Electric decades earlier. His work changed direction radically in 1948, after his Bell Lab colleagues invented the transistor. His job now was to develop a mass spectrograph to measure trace impurities in solids.

The first transistors were made from germanium, but the industry soon targeted silicon, with its lower cost and potential for developing a good oxide layer, as a replacement. At Bell Labs Hannay led both chemical and physical aspects of the silicon program. To avoid any contact between the silicon and any substance that might contribute impurities, his group devised the zone-refining method of growing silicon crystals in a vacuum, relying on mere surface tension to suspend the crystals. This method is used today to make substrates

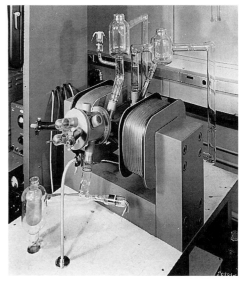

The mass spectrograph that N. Bruce Hannay developed for analyzing solids. Courtesy Bell Telephone Laboratories.

Four early single silicon crystals—the three top ones grown by the Czochralski ("pulling") method, the bottom one by the "floating zone" method. Courtesy Bell Telephone Laboratories.

for integrated circuit chips. In 1953 Bell Labs and Texas Instruments simultaneously announced the advent of the silicon transistor.

Through the 1950s Hannay and his colleagues investigated other semiconductors, including gallium arsenide—the preferred material for semiconductor lasers, which are the basis for optical communication systems today. They also made advances in the field of superconductors. Meanwhile Hannay rose in research management at Bell Labs and became the editor of the fundamental multivolume reference in the field, *Treatise on Solid State Chemistry*.

The next phase of the modern electronic revolution was the era of the computer chip. In 1962, about fifteen years after Hannay started at Bell Labs, Lubomyr T. Romankiw (Perkin Medal 1993) arrived at the Thomas J. Watson Research Center of IBM. Unlike Hannay, Romankiw came from a chemical engineering background—one that accustomed him to thinking in terms of tons of materials, not the micrograms demanded by the new electronics.

Romankiw grew up in the Ukraine, expecting to become a professional violinist, but in 1944 his family was forced to leave or face almost certain deportation to Siberia. His violin lessons were a casualty of the family's long westward trek via Europe to Canada. At the University of Alberta in Edmonton, where he soon enrolled, Romankiw chose chemical engineering as a major, influenced by a summer job at a metallurgical plant that recovered nickel, cobalt, and copper from sulfide ores by chemical means. There he also conceived the project that eventually brought him to graduate school at MIT: to develop a chemical means of extracting zinc from zinc sulfide ores instead of the traditional smelting process, which contributed to the acid rain already devastating Canadian forests. After reviewing his plan, the Department of Chemical Engineering at MIT referred him to the Department of Metallurgy and Materials. Although he designed a zinc-extracting process for his doctorate, his former employer in Canada had to a large extent anticipated his work. Romankiw was then at loose ends for a new research direction. In choosing IBM from a varied group of potential employers, he suddenly had to think on a wholly different scale. Or—as he edited the well-known IBM "THINK" sign issued to him in his new job—"THINK small."

In the early 1960s IBM researchers were at work on random access memory, or RAM, for computers, which they hoped to base on thin magnetic films incorporating complex circuitry. Romankiw was brought into the research purposely for his chemical background, and he and his group accordingly pursued an electrochemical means of depositing the necessary metal film. Meanwhile, a rival IBM group led by physicists and electrical engineers tried to achieve the same goal by using magnetic deposition techniques in a vacuum. Electroplating on such a small scale and with such detail and accuracy had never before been attempted and required specially designed techniques. Eventually Romankiw's group invented a tiny plating cell with a paddle to keep the plating bath circulating—"the Romankiw cell." They also used a photoresist mask as a template through which to electroplate the tiny circuitry—a

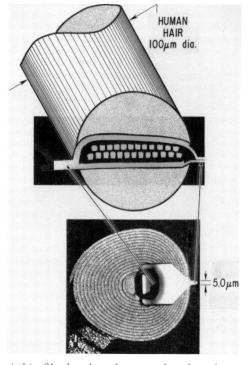

A thin film head used to record and read computer files stored magnetically on disks. Below is a top view of the electroplated magnetic yoke and the electroplated copper coil; above is a highly magnified cross-section of the head superimposed over a human hair. The chemical batch fabricating process, which combines optical lithography and electrochemical technology to mass-produce the heads, was pioneered by Lubomyr T. Romankiw and a group of his coworkers at the IBM Thomas J. Watson Research Center in Yorktown Heights, New York. Courtesy Lubomyr Romankiw and IBM Corporation.

process deemed impossible in standard electrochemical practice. Yet once Romankiw succeeded in converting management and the rival IBM group to his method, all his efforts suddenly seemed in vain when other researchers in the semiconductor industry succeeded in mass-producing the type of memory chips still in use in the 1990s.

But all that technical achievement did not go to waste. In 1968 a new challenge arose—how to reduce the size and cost of the heads used to record and read files stored magnetically on hard disks. These combination read-record heads then consisted of a core hand-wound with wire about the width of a human hair. For the writing process, a signal current flowing through the coil generated a magnetic field that magnetized selected parts of the hard disk; for reading, the previously magnetized regions of the disk induced electrical currents in the head that corresponded to the recording signal. Using the fundamental techniques developed in the memory project and with a great deal of cooperation from additional experts, Romankiw was able to plate magnets and conductor coils on thin films instead of wiring them. By 1980 the thin film heads were finally in mass production. They have since undergone several generations of improvement, and the density of information stored on hard disks—and consequently their capacity—has increased manyfold. Romankiw's techniques of electrodeposition have proven widely applicable in his own subsequent work and that of others in the microelectronics industry.

The Environment

The metaphor of the global village evokes not only communication but community, and the very computers, earth satellites, and other communication devices brought by the electronic revolution display to us the growing global impact of human activities. Chemistry too affects the environment we all share, and chemists too have shown concern for it. Yet so often what seems at first to be a step forward for civilization turns out to create problems that demand new solutions. Classic examples are Midgley's tetraethyl lead, which increased fuel efficiency in gasoline engines and made usable the petroleum fractions previously burned or vented into the atmosphere, and Freon, which replaced more hazardous compounds in refrigerators and air conditioners. Both, however, exacted their own price from the environment. Even clean water, better sanitation, and effective medicines—all based on chemical process industries—have a downside, in that by lowering overall mortality rates, they helped the world population skyrocket and thus put stress on global resources.

The Perkin and Chemical Industry medals, with their emphases on new products and processes, may seem unlikely mirrors to the environmental concerns of the larger chemical community, but their winners displayed that concern early on. Indeed the rise of the modern environmental movement in the 1960s spurred the award of numerous Perkin Medals to innovators of substitutes for earlier products found detrimental to the environment. And the recipi-

"A Proper Reception for King Cholera." Charles Chandler, with torch held high in this 1873 newspaper cartoon, led the drive for better sanitation in New York City as president of the Board of Health. Courtesy Chandler Museum, Columbia University.

ents of the Chemical Industry Medal, the managers in whose hands lie the direction of research and development—and the choice of products to manufacture—have in recent years often chosen corporate responsibility for the environment as the topic of their most urgent concern in their talks to fellow industrialists.

As in many other areas, Charles F. Chandler (Perkin Medal 1920) was a pioneer in his commitment to the environment. His work with New York City's Metropolitan Board of Health between 1867 and 1883 provided a model for laws and regulatory agencies nationwide, as he monitored food and drugs, provided free vaccinations, ensured the safety of milk supplies, brought clean water into the city, and enacted building codes with adequate provisions for indoor plumbing (which he personally designed with appropriate trapping systems). The Board of Health had to deal not only with problems created by large numbers of people living in unprecedented proximity, but also with those created by a burgeoning, if still chemically primitive, industry—the "nuisances" of noxious gases and acids discharged in sludge, as well as dangerous products like kerosene containing explosive naphtha fractions, or adulterated food, beverages, and cosmetics. When he received the Perkin Medal, Chandler—the man of a thousand endeavors—expressed surprise and delight, since he had never discovered an important process in industrial chemistry, though he had encouraged many a young chemist to pursue a career in innovation.

Two other Perkin medalists professionally active in sanitation and the pure food and drug movement before World War I were John Van Nostrand Dorr (Perkin Medal 1941) and Carl Shelley Miner (Perkin Medal 1949). The Dorr Company installed recovery and waste treatment plants for industries and water purification and sewage treatment plants for cities all over the country. Miner founded his consulting firm in 1906 to take advantage of the demand for chemical analyses generated by the landmark Pure Food and Drug Act of that year. Later the company became famous in the food processing industry for its useful advice on agricultural by-products.

Air pollution was also long recognized as a concomitant of industrialization. Indeed, the first instance of government regulation of the unintended by-products of industrial manufacturing was Britain's Alkali Acts of 1863, which sought to monitor the gaseous hydrochloric acid that spewed forth from plants manufacturing soda by the Leblanc process. The chemical innovation of passing the gas through water in a tower packed with coke was one of the first applications of chemical science to a threat to public health—just as the dissolved hydrochloric acid became a first among the useful by-products rescued in such environmental action.

Frederick G. Cottrell (Perkin Medal 1919), a Leipzig graduate and a professor at University of California, Berkeley, entered the pollution business via an acid—sulfuric acid. DuPont engaged him in 1906 as a consultant to a facility it had set up at Pinole, twenty miles north of Berkeley, to manufacture explosives and sulfuric acid by the contact process. Cottrell devised a means of re-

> **This chemical industry—_the_ basic industry—has the knowledge and technology to lead in every aspect of water management.**
>
> **—Leland I. Doan (Chemical Industry Medal 1964), "The Management of Water: A Challenge to the Chemical Industry"**

In the 1940s this Dorr waste-treatment plant at Calco Chemical Company, Bound Brook, New Jersey, processed 15 million gallons a day of organic and dye color waste. Courtesy American Cyanamid Company.

moving arsenic that was poisoning the catalyst by centrifuging the sulfuric acid mists. He then addressed the problem of precipitating the purified mist. He started experimenting with passing an electric charge to the mist globules, which then migrated to the oppositely charged electrode.

In 1907 Cottrell applied electrostatics to a different process, in connection with a successful court suit filed by Solano County, California, against the Selby Smelting and Lead Company, requiring the smelter to clean up its sulfurous smoke. The lead particles were filtered out by a "baghouse" fitted with 2,000 woolen bags, each thirty feet long, through which dust-laden gases were blown. Cottrell designed a precipitator to recover the sulfuric acid used in dissolving gold from the gold-and-silver alloy found in the lead. Later Cottrell installed similar equipment at a copper smelter and a cement factory, and developed a related electrostatic process for de-emulsifying oil. The fame of these operations—plus a presentation by Cottrell at the 1910 ACS meeting in San Francisco—let the world know that a major process for cleaning up the air was on the market. It was with profits from some of the associated patents that Cottrell set up the Research Corporation (see "Mortarboard and Lab Coat").

Walter A. Schmidt, president of Western Precipitation Company (left), and Research Corporation's founder, Frederick Gardner Cottrell, chief metallurgist of the Bureau of Mines when this photo was taken in 1916. Courtesy Research Corporation.

Later Environmentalists

One target of later Perkin medalists who worked on restoring and preserving our natural environment was automotive emissions. In his Perkin Medal address in 1959 Eugene J. Houdry, the pioneer of catalytic cracking, called for a thorough study of the effects of air contaminants on human health. A long-time devotee of sports cars, in the 1950s he founded a special company, Oxy-

The first experimental electrostatic precipitator at the University of California, Berkeley. Sulfuric acid was generated, then bubbled through water in a U-tube beneath an inverted glass bell jar. The inner walls of the jar, quickly wetted with acid, served as a collecting electrode and were connected to an induction coil (left), which acted as a step-up transformer. Left: With the current off, acid mist pours from the bell jar. Right: With the current on, the acid mist disappears. Courtesy Research Corporation.

Keeping water clean is a job for chemistry. Left: Today Nalco helps companies treat industrial wastes to remove solids and oils and lower "BOD" (biochemical oxygen demand) before discharge. Right: Nalco's pollution-control surveillance programs incorporate monitoring plant effluent and conducting river-profile studies. Both courtesy Nalco Chemical Company.

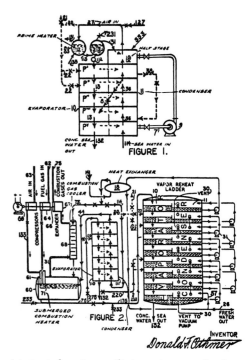

A patent drawing outlining a process developed by Donald Othmer for making pure water from sea water. Courtesy Donald F. Othmer.

Catalyst, to develop catalytic converters for gasoline engines—an idea ahead of its time. But until some way could be found to get the lead out of gasoline, it would poison the catalyst. That problem was solved by John H. Sinfelt (Perkin Medal 1984), who devised a process to produce unleaded high-octane gasoline—and eliminated lead emissions. Donald Othmer (Perkin Medal 1978), whose career started in the days when organic chemicals came largely from wood and coal, has long advocated alternative fuels. His many patents include one for a process to prepare a methanol fuel for powering buses and automobiles, and one for making drinkable water from sea water.

Two miracle products, pesticides and synthetic detergents, were belatedly discovered to be so chemically stable that they accumulated in water supplies and the food chain. An early harbinger of challenges ahead were detergent suds that turned out to be extraordinarily long-lived, clogging up wastewater treatment facilities and reappearing in brooks and streams. The detergent industry quickly found a solution in less highly branched molecules that natural microorganisms can digest. The particular ones needed had never been produced on a commercial scale, but James Roth (Perkin Medal 1988) at Monsanto spearheaded the effort that produced a commercial process, using a novel and efficient direct dehydrogenation of normal (straight-chain) paraffins to form olefins that can link to the benzene ring of a detergent molecule.

In her eye-opening book *Silent Spring* (1962) the biologist Rachel Carson explained how DDT, the wonder insecticide that did so much to check mos-

Monsanto's John Franz, who in 1970 discovered the environmentally friendly herbicide glyphosate, holds a flower in a beaker—clearly a form of plant life he would like to protect from the encroachment of weeds. Courtesy Monsanto Company.

Spraying Rodeo, a variant of Roundup designed for wetlands. Courtesy Monsanto Company.

quito-borne diseases worldwide, moved up the food chain, causing problems such as bird eggs that were too thin to sustain developing chicks. Without birds, spring would indeed be silent. Since the 1960s chemists have developed new generations of pesticides that break down more rapidly instead of accumulating in soil and water. Herbicides to help ensure an abundant food supply for a growing world population without harming the environment are the contribution of two recent Perkin Medalists. John E. Franz (Perkin Medal 1990) of Monsanto was educated as an organic chemist at the Universities of Illinois and Minnesota. After a dozen years with the company he was transferred to Monsanto's agricultural unit in 1967, where he had to teach himself plant physiology and biochemistry. In his new assignment he took up a project that other investigators had dropped—finding a herbicide effective against both perennial and annual weeds. Two compounds were known to be weakly active in this regard, but nine years of trying various analogues failed to produce a more effective herbicide. Franz then tried the analogue route for a year and finally hypothesized—falsely, as it turned out—that the weakly effective compounds were metabolized in the plant into herbicidal agents. On the basis of this hypothesis, he screened possible metabolites and came across glyphosate, which turned out to be broadly effective as a herbicide and without toxic effects on mammals, birds, fish, insects, and most bacteria. Only later was the activity and selectivity of the product (sold as "Roundup") correctly explained; the glyphosate inhibited the formation of an enzyme found only in plants.

Marinus Los (Perkin Medal 1994) of American Cyanamid discovered another class of herbicidal compounds that are not toxic to humans and animals—

imidazolinones. After graduating with honors from Edinburgh University, he joined American Cyanamid as a research chemist in 1960. His major breakthrough came after fourteen years of research. The new herbicide, introduced in 1985, is so effective that it can be applied in ounces per acre, and its use has reduced the amount of herbicide applied annually by a total of 70 million pounds. The development of these highly selective and effective compounds allowed farmers to increase yields while preserving our land, water, and wildlife.

All the efforts that the chemical industry has made in recent years to achieve environmentally sustainable development could not of course be represented in Perkin Medals, in part because these initiatives were often incremental improvements or even decisions to halt production. The industry leaders selected for the Chemical Industry Medal have made their own contributions on these lines. To cite just two: W. H. Clark (Chemical Industry Medal 1993), chairman and CEO of Nalco, a company with expertise in all aspects of water treatment needed by industry and by consumers, helped make this multipronged know-how available in all corners of the globe. And after chlorofluorocarbons were shown to be harmful to the ozone layer, DuPont CEO Richard E. Heckert (Chemical Industry Medal 1989) decided that DuPont would cease producing them—and vigorously pursued the search for environmentally acceptable replacements.

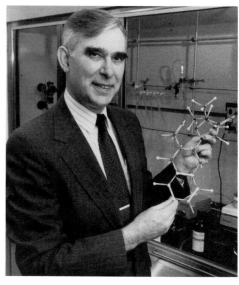

Marinus Los, director of crop science discovery at American Cyanamid's agricultural research laboratory in Princeton, holds a molecular model of one of the environmentally friendly imidazolinone herbicides first marketed in 1985. Courtesy American Cyanamid Company.

Imidazolinone herbicides can be applied in extremely low quantities to control weeds in soybeans (as here) or cereal crops. The products are nontoxic to humans and wildlife and do not contaminate ground water because they remain in the top layers of the soil until degraded. Courtesy American Cyanamid Company.

In Conclusion

The companies that will be most profitable in the long run will be those that serve society best. Society will reward those that help unclog our highways, rebuild and revitalize our cities, cleanse our streams, and conquer poverty and disease, not those whose pursuit of the dollar blinds them to such needs.

—Charles B. McCoy, Chemical Industry Medal Address (1969)

Chemistry and chemical technology have been at the heart of the revolutionary developments of the twentieth century—the automobile and airplane; movies, radio, and television; central heating, indoor plumbing, and air conditioning; plastics, synthetic fibers, and a host of other new materials that serve in myriad ways; the silicon chip and the computerization of virtually everything; astounding increases in agricultural productivity; and the modern pharmaceutical assault on disease.

The careers of the winners of the Perkin Medal reflect how, though the goals of technological research have remained the same since Perkin's time, the process of innovation has changed dramatically. At the turn of this century individual inventors, entrepreneurs, and consultants could assemble the necessary knowledge, skill, and capital to innovate successfully. Opportunities abounded as an ebullient, expanding science of chemistry gave the early Perkin medalists deeper insight into existing technologies and suggested novel processes and products. After a new idea proved sound at the laboratory bench, the innovator often became an engineer as well as a chemist, in order to scale up the process. But by starting small and expanding slowly, entrepreneurs could manage the technical and financial requirements of their businesses.

As the scale of chemical enterprise grew still further and as chemistry and chemical engineering became more sophisticated disciplines, the frontier of innovation passed to the corporation, where research directors and corporate managers risked millions of dollars on new products and processes. The orchestration of the various talents and skills needed for innovation required conductors with great vision, perseverance, powers of persuasion, and ability to lead others into the unknown. Effective management became as critical to innovation as good science and technology.

Chemistry solving the problems of industrialization: Eugene Houdry in 1953, holding a small catalytic converter (see "In a Global Village"). Courtesy Sun Company.

The creation of the Chemical Industry Medal to stand beside the Perkin Medal signalled the emergence of chemical administration as a vital, special field. The ranks of Chemical Industry medal winners are full of individuals who became highly proficient in the difficult, subtle arts of management. In his 1987 Medal address Edwin C. Holmer of Exxon Chemical voiced a sentiment such managers share, when he emphasized "the human chemistry of the chemical industry," the good old-fashioned teamwork that is essential for success in today's complex technological and competitive business world.

Innovation within corporations nonetheless still depends on the creative chemists and chemical engineers who make the initial breakthroughs. In the words of one highly creative chemical engineer, Ralph Landau: "There is no doubt that the increasing complexity and scale of modern technology make the role of the individual less likely than ever to be decisive as sole or principal technological innovator, which only emphasizes that nevertheless the creative individual must not be overwhelmed or submerged by excessively rigid or complex institutional structures." Even the passing of that heroic American figure—the individual innovator—can be exaggerated. In the past twenty years Edwin Land (instant photography), Carl Djerassi (pharmaceuticals), Donald Othmer (distillation), and Landau himself have won Perkin Medals for their efforts in entrepreneurial settings.

Outstanding records of innovation in chemicals soon made it one of America's most successful high-technology industries. In 1950 *Fortune* magazine labeled the twentieth century the chemical century. Yet despite such proclamations the industry churned along quietly and efficiently for decades, largely unnoticed by the general public. One notable exception may be prudent investors who benefit from the industry's consistent growth and profitability: it remains true today that the American chemical industry is highly competitive in international markets and generates a significant trade surplus for the United States.

The impressive record of innovation and profitability did not prevent two different issues from making the chemical industry a focus for debate and discussion in the media and in government in the 1960s and 1970s. These issues were the transformation of sensibility about the environment that found its focus in Earth Day in 1970, and the dramatic threat posed to American lifestyles by the energy crisis and the international politics of petroleum. The legacy of these two issues means that today's executives have not only to run their respective companies but also to play an ever-increasing role in the public and political spheres.

The remarks of recent Chemical Industry Medal winners reflect the industry's commitment to confronting problems in these areas and contributing to their solution. As early as 1970, William H. Lycan, vice president of research at Johnson & Johnson, on accepting his award chose to highlight how many of the really serious problems in the world were environmental: "The management of water supplies, development of energy sources, population, and to

Innovation creates a chemical lifesaver: fire-resistant suit for firefighters made of poly-benzimidazole (PBI). Carl Marvel discovered PBI in 1983, when he was searching for a more heat-resistant material for truck tires (see "A Symphony of Synthetics"). Courtesy Hoechst Celanese Corporation.

some extent pollution—those are the key problems. . . . Scientists and particularly industrialists have to become more vocal in expressing solutions to [what] are essentially technical problems, [problems] that won't be solved by any amount of jawboning or wringing of hands." Lycan's remarks proved prescient both in terms of the emerging agenda and the chemical industry's attitude.

Later medalists repeatedly stressed the need for the industry to have more effective dialogue with the press, politicians, and the general public. In 1985 Monsanto's Louis Fernandez argued:

> We have to spend more time communicating with the people in the media, helping them to understand what the industry is doing, . . . making them comfortable with what we are doing, being honest with them when we're doing something wrong and when we have made mistakes. Then we have to supplement that with what I think is a long-range program of education, which ought to be perceived as a broader societal need—that is, to get the American public to become more comfortable with technology.

The 1992 medalist, H. Eugene McBrayer, successor to Edwin Holmer as president of Exxon Chemical, stressed that the chemical industry must help improve the quality of science education in America. It must also improve the industry's credibility, something that Responsible Care—a program launched by the Chemical Manufacturers Association—has already begun to do.

Having a scientifically literate public, one comfortable with chemistry, is important not only for dealing with domestic issues but also for insuring America's strength in the global economy. Because most U.S. chemical companies operate internationally, trade issues have become yet another of the chief executive officer's necessary fields of expertise. For instance, the 1993 medalist, W. H. Clark, chairman and CEO of Nalco Chemical and a major spokesman for NAFTA (North American Free Trade Agreement), used the bully pulpit of his medal address to emphasize the intricacies and the importance of international trade negotiations. A worldwide market presents opportunities as well as challenges. Thus Dexter F. Baker (Chemical Industry Medal 1991) of Air Products and Chemicals emphasized that access to international markets is needed to recover the increasing costs of research and development. Before succeeding Edward Donley (Chemical Industry Medal 1980) as the head of Air Products, Baker developed a major European business for the company in the late 1950s and 1960s.

The globalization of chemical activity has intensified competition, putting more pressure on the traditional realms of research, manufacturing, and marketing at the same time that the repertoire of skills needed by a CEO has expanded so greatly. If the Chemical Industry Medal winner of today is the manager of an enterprise, a public spokesman, and a global strategist, he is also someone who understands the importance of the Perkin tradition and who sees innovation as the key to long-term success.

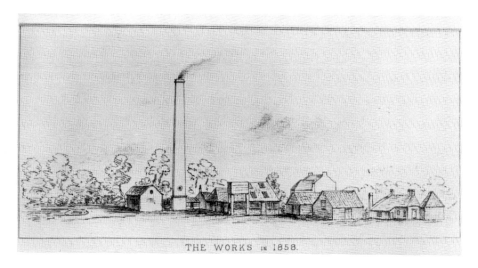

THE WORKS in 1858.

Where it all began: the Perkin works in 1858. Courtesy the Edelstein Collection, Hebrew University, Jerusalem.

Linking management and innovation, George J. Sella, Jr., won the Chemical Industry Medal in 1990 for his dramatic expansion of research and development activities at American Cyanamid during the 1980s. His predecessor, James G. Affleck, similarly emphasized the need for more research in his 1984 medal address. Affleck and Sella were convinced that, as in the past, innovation will be the key to success in the future. This recurrent commitment to pioneering research was heralded by the 1986 medalist, DuPont's Edward Jefferson as well. Jefferson pointed to the historical connections between science and industry: "If you look down the years you'll find that some of the most creative basic scientists had a high interest in seeing the benefits of their basic work translated into things of value for their society. In fact, the areas in which they worked were often very much influenced by what society needed."

Jefferson's remark accurately reflects the long tradition in the chemical industry of combining theory and practice, science and engineering, technology and business to improve the lot of humankind. While this central tradition remains unchanged, the methods of achieving it have become much more complex over the years. Environmental, educational, political, and trade issues have joined those of innovation and creativity. Current executives may well envy the simpler world of their predecessors. The centrality of the chemical industry to the modern world means that it will continue to offer challenging, rewarding careers to individuals of talent and motivation. The chemical industry, and the dedicated individuals who bring it to life, will surely be central to the story of the twenty-first century—the story of the continuing effort to provide a decent life for all the inhabitants of our global village.

What would Henry Perkin's reactions be today if he came back to see how well the chemical industry had pursued and built on his landmark invention? The answer must be very positive. . . . Perkin must be smiling with great satisfaction that we have indeed kept the faith.

—Vincent L. Gregory, Chemical Industry Medal address, 1988

American Section

Chairmen, 1894–1994

1894–1896
ALFRED H. MASON

1896–1898
CHARLES F. CHANDLER

1898–1900
THOMAS J. PARKER

1900–1902
CLIFFORD RICHARDSON

1902–1904
VIRGIL COBLENTZ

1904–1906
RUSSELL W. MOORE

1906–1908
GEORGE C. STONE

1908–1910
MAXIMILIAN TOCH

1910–1911
CHARLES F. McKENNA

1911–1913
MILTON C. WHITAKER

1913–1915
GUSTAVE W. THOMPSON

1915–1916
WILLIAM M. GROSVENOR

1916–1918
JEROME ALEXANDER

1918–1920
CHARLES E. SHOLES

1920–1922
SUMNER R. CHURCH

1922–1924
RALPH H. McKEE

1924–1926
HARLAN S. MINER

1926–1928
LAWRENCE V. REDMAN

1928–1930
CHARLES A. LUNN

1930–1931
DANIEL D. JACKSON

1931–1932
ALLEN ROGERS

1932–1934
ALBERT E. MARSHALL

1934–1936
ROBERT J. MOORE

1936–1938
JAMES G. VAIL

1938–1940
WALLACE P. COHOE

1940–1942
LINCOLN T. WORK

1942–1944
FOSTER D. SNELL

1944–1945
NORMAN A. SHEPARD

1945–1946
FRANCIS J. CURTIS

1946–1947
SIDNEY D. KIRKPATRICK

1947–1948
CYRIL S. KIMBALL

1948–1949
ARCHIE J. WEITH

1949–1951
GUSTAVUS J. ESSELEN

1951–1952
ROBERT C. SWAIN

1952–1953
HARRY B. McCLURE

1953–1954
LAUREN B. HITCHCOCK

1954–1955
CLIFFORD F. RASSWEILER

1955–1956
RAYMOND STEVENS

1956–1957
MONROE E. SPAGHT

1957–1958
WILLIAM H. BOWMAN

1958–1959
WILLIAM E. HANFORD

1959–1960
E. DUER REEVES

1960–1961
WILLIAM H. LYCAN

1961–1962
ROBERT W. CAIRNS

1962–1963
CECIL W. HUMPHREYS

1963–1964
DONALD B. BENEDICT

1964–1965
SAMUEL LENHER

1965–1966
GLENN A. NESTY

1966–1967
MAX TISHLER

1967–1968
RICHARD O. ROBLIN

1968–1969
JESSE WERNER

1969–1970
RICHARD R. MESSING

1970–1971
FREDERIC A.L. HOLLOWAY

1971–1972
WILLIAM J. HAINES

1972–1973
NOLAN B. SOMMER

1973–1974
EDWARD R. KANE

1974–1975
LUTHER S. ROEHM

1975–1976
WERNER C. BROWN

1976–1977
EDWARD J. GOETT

1977–1978
VINCENT L. GREGORY, JR.

1978–1979
MONTE C. THRODAHL

1979–1980
RALPH LANDAU

1980–1981
EDWARD G. JEFFERSON

1981–1982
H. BARCLAY MORLEY

1982–1983
CARLYLE G. CALDWELL

1983–1984
WARREN M. ANDERSON

1984–1985
EDWIN C. HOLMER

1985–1986
LOUIS FERNANDEZ

1986–1987
GEORGE J. SELLA, JR.

1987–1988
RICHARD E. HECKERT

1988–1989
DEXTER F. BAKER

1989–1990
JAMES R. STREET

1990–1991
L. JOHN POLITE, JR.

1991–1992
W. H. CLARK

1992–1993
KEITH R. McKENNON

1993–1994
J. LAWRENCE WILSON

1994–1995
HAROLD A. SORGENTI

Perkin Medalists, 1906–1994

1906
SIR WILLIAM H. PERKIN

1908
J.B. FRANCIS HERRESHOFF

1909
ARNO BEHR

1910
EDWARD G. ACHESON

1911
CHARLES M. HALL

1912
HERMAN FRASCH

1913
JAMES GAYLEY

1914
JOHN W. HYATT

1915
EDWARD WESTON

1916
LEO H. BAEKELAND

1917
ERNST TWITCHELL

1918
AUGUSTE J. ROSSI

1919
FREDERICK G. COTTRELL

1920
CHARLES F. CHANDLER

1921
WILLIS R. WHITNEY

1922
WILLIAM M. BURTON

1923
MILTON C. WHITAKER

1924
FREDERICK M. BECKET

1925
HUGH K. MOORE

1926
RICHARD B. MOORE

1927
JOHN E. TEEPLE

1928
IRVING LANGMUIR

1929
EUGENE C. SULLIVAN

1930
HERBERT H. DOW

1931
ARTHUR D. LITTLE

1932
CHARLES F. BURGESS

1933
GEORGE OENSLAGER

1934
COLIN G. FINK

1935
GEORGE O. CURME, JR.

1936
WARREN K. LEWIS

1937
THOMAS MIDGLEY, JR.

1938
FRANK J. TONE

1939
WALTER S. LANDIS

1940
CHARLES M.S. STINE

1941
JOHN V. N. DORR

1942
MARTIN ITTNER

1943
ROBERT E. WILSON

1944
GASTON F. DUBOIS

1945
ELMER K. BOLTON

1946
FRANCIS C. FRARY

1947
ROBERT R. WILLIAMS

1948
CLARENCE W. BALKE

1949
CARL S. MINER

1950
EGER V. MURPHREE

1951
HENRY HOWARD

1952
ROBERT M. BURNS

1953
CHARLES A. THOMAS

1954
ROGER ADAMS

1955
ROGER WILLIAMS

1956
EDGAR C. BRITTON

1957
GLENN T. SEABORG

1958
WILLIAM J. KROLL

1959
EUGENE J. HOUDRY

1960
KARL FOLKERS

1961
CARL F. PRUTTON

1962
EUGENE G. ROCHOW

1963
WILLIAM O. BAKER

1964
WILLIAM J. SPARKS

1965
CARL S. MARVEL

1966
MANSON BENEDICT

1967
VLADIMIR HAENSEL

1968
HENRY B. HASS

1969
ROBERT W. CAIRNS

1970
MILTON HARRIS

1971 JAMES FRANKLIN HYDE	**1977** PAUL J. FLORY	**1983** N. BRUCE HANNAY	**1989** FREDERICK J. KAROL
1972 ROBERT BURNS MACMULLIN	**1978** DONALD F. OTHMER	**1984** JOHN H. SINFELT	**1990** JOHN E. FRANZ
1973 THEODORE L. CAIRNS	**1979** JAMES D. IDOL, JR.	**1985** PAUL B. WEISZ	**1991** MIGUEL A. ONDETTI
1974 EDWIN H. LAND	**1980** HERMAN F. MARK	**1986** PETER REGNA	**1992** EDITH M. FLANIGEN
1975 CARL DJERASSI	**1981** RALPH LANDAU	**1987** J. PAUL HOGAN & ROBERT L. BANKS	**1993** LUBOMYR T. ROMANKIW
1976 LEWIS H. SARETT	**1982** HERBERT C. BROWN	**1988** JAMES F. ROTH	**1994** MARINUS LOS

Chemical Industry Medalists, 1933–1994

1933
JAMES G. VAIL

1934
FLOYD J. METZGER

1935
EDWARD R. WEIDLEIN

1936
WALTER S. LANDIS

1937
EVAN J. CRANE

1938
JOHN V. N. DORR

1939
ROBERT E. WILSON

1941
ELMER K. BOLTON

1942
HARRISON E. HOWE

1943
JOHN J. GREBE

1944
BRADLEY DEWEY

1945
SIDNEY D. KIRKPATRICK

1946
WILLARD H. DOW

1947
GEORGE W. MERCK

1948
JAMES A. RAFFERTY

1949
WILLIAM B. BELL

1950
WILLIAM M. RAND

1951
ERNEST W. REID

1952
CRAWFORD H. GREENEWALT

1953
CHARLES S. MUNSON

1954
ERNEST H. VOLWILER

1955
JOSEPH G. DAVIDSON

1956
R. LINDLEY MURRAY

1957
CLIFFORD F. RASSWEILER

1958
FRED J. EMMERICH

1959
HARRY B. McCLURE

1960
HANS STAUFFER

1961
WILLIAM E. HANFORD

1962
KENNETH H. KLIPSTEIN

1963
MAX TISHLER

1964
LELAND I. DOAN

1965
RALPH CONNOR

1966
MONROE E. SPAGHT

1967
CHESTER M. BROWN

1968
HAROLD W. FISHER

1969
CHARLES B. McCOY

1970
WILLIAM H. LYCAN

1971
CARROLL A. HOCHWALT

1972
JESSE WERNER

1973
RALPH LANDAU

1974
CARL A. GERSTACKER

1975
LEONARD P. POOL

1976
HAROLD E. THAYER

1977
F. PERRY WILSON

1978
JACK B. ST. CLAIR

1979
IRVING SHAPIRO

1980
EDWARD DONLEY

1981
THOMAS W. MASTIN

1982
H. BARCLAY MORLEY

1983
PAUL F. OREFFICE

1984
JAMES AFFLECK

1985
LOUIS FERNANDEZ

1986
EDWARD G. JEFFERSON

1987
EDWIN C. HOLMER

1988
VINCENT L. GREGORY

1989
RICHARD E. HECKERT

1990
GEORGE J. SELLA, JR.

1991
DEXTER F. BAKER

1992
H. EUGENE McBRAYER

1993
W. H. CLARK

1994
KEITH R. McKENNON

For Further Reading

Fred Aftalion. *The History of the International Chemical Industry.* Philadelphia: University of Pennsylvania Press, 1991.

Theodor Benfey, editor. *Introducing the Chemical Sciences: A CHF Reading List.* Philadelphia: Chemical Heritage Foundation, 1994.

Clark K. Colton, editor. *Perspectives in Chemical Engineering: Research and Education.* Boston: Academic Press, 1991.

William F. Furter, editor. *A Century of Chemical Engineering.* New York: Plenum, 1982.

——, editor. *History of Chemical Engineering.* Washington, D.C.: American Chemical Society, 1980.

L.F. Haber. *The Chemical Industry during the Nineteenth Century.* Oxford: Oxford University Press, 1971.

——. *The Chemical Industry, 1900–1930: International Growth and Technological Change.* Oxford: Clarendon Press, 1971.

Williams Haynes. *American Chemical Industry: A History.* 6 vols. New York: Van Nostrand, 1945-1954.

——. *Men, Money, and Molecules: The Story of Chemical Business.* New York: Doubleday, Doran, 1936.

John A. Heitmann; David J. Rhees. *Scaling Up: Science, Engineering, and the American Chemical Industry.* Philadelphia: Beckman Center for the History of Chemistry, Chemical Heritage Foundation, 1984.

Gino J. Marco; Robert M. Hollingworth; William Durham, editors. *Silent Spring Revisited.* Washington, D.C.: American Chemical Society, 1987.

Wyndham D. Miles, editor. *American Chemists and Chemical Engineers.* Washington, D.C.: American Chemical Society, 1976.

Wyndham D. Miles; Robert F. Gould, editors. *American Chemists and Chemical Engineers.* Volume 2. Guilford, Conn.: Gould Books, 1994.

Tom Mahoney. *The Merchants of Life: An Account of the American Pharmaceutical Industry.* New York: Harper, 1959.

John Mann. *Murder, Magic, and Medicine.* New York: Oxford University Press, 1992.

Peter J.T. Morris. *Polymer Pioneers: A Popular History of the Science and Technology of Large Molecules.* Philadelphia: Beckman Center for the History of Chemistry, Chemical Heritage Foundation, 1986.

——. *The United States Synthetic Rubber Research Project.* Philadelphia: University of Pennsylvania Press, 1989.

Nicholas A. Peppas, editor. *One Hundred Years of Chemical Engineering.* Dordrecht/Boston: Kluwer Academic Publishers, 1989.

Terry S. Reynolds. "Defining Professional Boundaries: Chemical Engineering in the Early 20th Century." *Technology and Culture* 27 (1986), 694–716. Reprinted in *The Engineer in America,* edited by Reynolds. Chicago: University of Chicago Press, 1991.

Nathan Rosenberg; Ralph Landau; David C. Mowery, editors. *Technology and the Wealth of Nations.* Stanford: Stanford University Press, 1992.

Raymond B. Seymour, editor. *History of Polymer Science and Technology.* New York/Basel: Marcel Dekker, 1982.

Edwin E. Slosson. *Creative Chemistry* (1919). New edition. New York: Appleton-Century, 1930.

John Kenly Smith. "The Evolution of the Chemical Industry: A Technological Perspective." In *The Chemical Sciences in the Modern World,* edited by Seymour H. Mauskopf. Philadelphia: University of Pennsylvania Press, 1993.

Peter H. Spitz. *Petrochemicals: The Rise of an Industry.* New York: John Wiley & Sons, 1988.

Jeffrey L. Sturchio, editor. *Corporate History and the Chemical Industry: A Resource Guide.* Philadelphia: Beckman Center for the History of Chemistry, Chemical Heritage Foundation, 1985.

John P. Swann. *Academic Scientists and the Pharmaceutical Industry: Cooperative Research in Twentieth-Century America.* Baltimore: Johns Hopkins University Press, 1988.